CELL BIOLOGY RESEARCH PROGRESS

TISSUE ENGINEERING

FUNDAMENTALS, TECHNIQUES AND APPLICATIONS

Cell Biology Research Progress

Additional books in this series can be found on Nova's website
under the Series tab.

Additional e-books in this series can be found on Nova's website
under the e-book tab.

Stem Cells and Regenerative Medicine

Additional books in this series can be found on Nova's website
under the Series tab.

Additional e-books in this series can be found on Nova's website
under the e-book tab.

CELL BIOLOGY RESEARCH PROGRESS

TISSUE ENGINEERING

FUNDAMENTALS, TECHNIQUES AND APPLICATIONS

TAKESHI ISHIKAWA
AND
MASAYOSHI YAMAMOTO
EDITORS

NOVA BIOMEDICAL
New York

For permission to use material from this book please contact us:
Telephone 631-231-7269; Fax 631-231-8175
Web Site: http://www.novapublishers.com

Additional color graphics may be available in the e-book version of this book.

Library of Congress Cataloging-in-Publication Data
Tissue engineering : fundamentals, techniques and applications / editors, Takeshi Ishikawa and Masayoshi Yamamoto.
p. cm.
Includes index.
ISBN 978-1-62257-062-1 (hardcover)
1. Tissue engineering. 2. Liver--Transplantation. I. Ishikawa, Takeshi. II. Yamamoto, Masayuki, 1954-
R857.T55T553 2012
616.02'7--dc23

2012018471

Published by Nova Science Publishers, Inc. † New York

CONTENTS

PREFACE

Tissue engineering and regenerative medicine uses a combination of cells, scaffolding and bioreactive factors to treat a variety of pathological conditions and has become a treatment option for many adult diseases. In this book, the authors present current research from across the globe, in the study of the fundamentals, techniques and applications of tissue engineering. Topics discussed in this compilation include the characterization of liver organogenesis and fetal and adult stem/progenitor cells; in vitro biological activity of double and triple component system scaffolds in bone tissue engineering; stretching bioreactors for dynamic engineering of muscle tissues; adipose-derived stem cells and their application in tissue engineering; regenerative medicine and tissue engineering for congenital birth defects.

Chapter 1 - Liver transplantation is the standard of care for many end-stage liver diseases. But donor organs are scarce, and some people succumb to liver failure before a donor is found. Liver regenerative medicine is a special interdisciplinary field of medicine focused on the development of new therapies incorporating stem cells, gene therapy, and engineered tissues in order to repair or replace the damaged organ. This review considers the emerging progress achieved in the hepatic regenerative medicine within the last decade. The review starts from the characterization of liver organogenesis, fetal and adult stem/progenitor cells. Then, applications of primary hepatocytes, embryonic and adult (mesenchymal, hematopoetic, and induced pluripotent) stem cells in cell therapy of liver diseases are considered. Current advances and challenges in producing mature hepatocytes from stem/progenitor cells are discussed. A section about hepatic tissue engineering includes consideration of synthetic and natural biomaterials in engineering scaffolds, strategies and achievements in the development of 3D bioactive matrices and 3D hepatocyte cultures, liver microengineering, generating bioartificial liver and prospects in fabrication of the bioengineered liver.

Chapter 2 - Tissue engineering is a vast and emerging area of research in the recent years. It has been significantly reduced the secondary operation and transmissible disease as well as the huge amount of expensive compared with allograft and allograft techniques. Bone is made of hydroxyapatite as well as collagen as a major portion. The great interest has been given by the researchers in the preparation of artificial bone like materials.

For this, scaffolds are made by natural based polymers cellulose, chitin, chitosan, alginic acid and amylopectin are commonly employed. The addition of bioceramics (nano hydroxylapatite and tricalcium phosphate) and other promising materials (carbon nanotube, grapheme and boron nitride nanotubes) with biopolymers are great interest to accomplish the goal in the preparation of bone like material.

In this chapter, the preparation, characterization and *in vitro* biological activity of double and triple component system scaffolds have been explored for bone tissue engineering application.

Chapter 3 - *Objectives:* Mechanical stimulation of engineered tissue holds great promise for the development of functionally structured tissues. In particular, mimicking the mechanical properties of native heart and the loading conditions of the cardiac tissue in vitro relies on dynamic, three dimensional culture systems that simulate the anatomical geometry and phasic contraction of cardiac left ventricles (LV). Chapter 3 reports the study on the development and validation of a new device composed of a LV shaped muscle biograft built on electrospun fibers submitted to circumferential stretch.

Methods: Custom made silicone left ventricle (LV) shaped bulbs were produced with a water-soluble wax mold and covered with an electrospun poly(ε-caprolactone) (PCL) micron-scaled fibrous matrix. Pump controlled volumetric changes induced bulb enlargement and resulted in matrix stretching. Spatial characterization of the stretch was analyzed using a GOM 3D digitizer and GOM ARAMIS software. C_2C_{12} cells were seeded on the matrix and cultured for 1 week under mechanical stimulation. Cyclic (1Hz) and ramp (cycles of 24h stretch /24h rest) stretches were applied. Cellular distribution was investigated by DNA quantification and tissue thickness evaluated using 3D confocal microscopy.

Results: A gradient of surface strains was characterized from the base to the apex of the bulb. When inflated, the apex showed a linear increase in the strain from 4 to 24%. Meanwhile, the increase of strain at the base was limited and ranged from 0% to 5%. Dynamic culture induced variations in cell density and tissue thickness. This effect appeared to be dependent on the applied stretch amplitude.

Conclusion: The authors provide evidence that the new device composed of a LV shaped bulb covered with a microfibrous matrix is promising for structured muscle tissue generation and will further generate a broad interest from researchers investigating the effect of anisotropic and circumferential stretch on cardiac stem cells and cardiac physiopathology.

Chapter 4 - Tissue engineered constructs typically comprise of a cellular and biomaterial component. The latter takes on the role of a scaffold, to permit aggregation and organization of 3-dimensional extracellular matrix or tissue produced by the cells. However, it is clear that in areas of tissue engineering where tissue defects, voids, gaps or spaces need to be filled, proper *in situ* tissue formation will require sufficient spatial and temporal retention of the engineered construct at the repair site. Thus far, limited efforts have focused on this important design consideration required for the proper implantation of tissue engineered materials. Yet, the integration of the biomaterial with surrounding tissue is critical in maximizing long-term functionality and success. This can be achieved by one of 4 fixation methods: i) chemical, ii) mechanical, iii) biological or iv) some combination of i) through iii). Importantly, it is critical that the methodologies applied do not interfere with the tissue repair and regeneration processes. In this chapter, strategies are reviewed for enhanced integration between biomaterial and native tissues, the ultimate goal being that *in situ* engineered tissue formation can be augmented. Emphasis is placed in the area of cartilage tissue engineering wherein strong integration between engineered cartilage and subchondral bone is required.

Chapter 5 - Regenerative medicine is emerging as a rapidly evolving field of research and therapeutics, and stem cells hold great promise for future translational research and widespread clinical application. Adipose-derived stem cells (ASCs) are as capable as bone marrow-derived stem cells (BMSCs) of differentiating into multiple lineages, such as bone, cartilage, tendons, nerves and fat, when cultivated under lineage-specific conditions.

Furthermore, ASCs appear to have comparable properties to BMSCs as well as favorable angiogenic and anti-inflammatory secretion profiles. As human adipose tissue can be obtained easily in large quantities under local anesthesia with little patient discomfort, it may thus provide an alternative source of stem cells for mesenchymal tissue regeneration and engineering.

Chapter 5 describes the methods used the laboratory and in the literature for the isolation and expansion of ASCs. In addition, presented is the potential application of ASCs in tissue engineering, particularly in the regeneration of fat, bone and cartilage and comment on the progress of this approach and future trends in the field. The fast expanding field of basic and clinical research into ASC biology, pathophysiology, and potential therapeutic applications promises further exciting discoveries in the near future.

Chapter 6 - Tissue engineering and regenerative medicine (TERM) uses a combination of cells, scaffolding and bioreactive factors to treat a variety of pathologic conditions and has become a treatment option for many adult diseases. The use of TERM to treat congenital birth defects has not been well established. However, for the following reasons, it has the potential to be even more successful in pediatric applications than with adults. First, TERM is minimally invasive and a gentle treatment strategy for small infants with congenital birth defects. Second, tissue regeneration in infants is much more vigorous than that in adults because the number of stem and progenitor cells as well as the proliferative capacity of the latter is greater in infants. Third, regenerated tissues are expected to grow with age. This is an important consideration in the treatment of infants. Furthermore, there is no need to take immunosuppressive drugs during growth and maturation of the child. Thus, the use of TERM in infants should be particularly advantageous. The authors investigated the possibility of using TERM in treatments for congenital soft tissue defects that currently require surgical repair, such as esophageal atresia, some bladder anomalies that require augmentation, diaphragmatic hernia, and imperforate anus. Also, treatment strategies using TERM for these defects and the results obtained with animal models are presented. These include *in situ* regeneration of esophagus using gastric acellular matrix, *ex vivo* construction of the urinary bladder wall, diaphragmatic repair using a combination of scaffolds and marrow-derived mesenchymal stem cells, and a novel therapy for fecal incontinence after surgical repair of imperforate anus by myoblast transplantation. Finally, the study is introduced on the potential use of stem cells derived from the amnion as a novel stem cell source for congenital and perinatal diseases. The current status and future prospects of TERM in the treatment of congenital birth defects are reviewed and discussed.

Chapter 7 - The reoccurring problems faced by the healthcare system today are the loss or failure of tissues or organs. Repairing damaged or diseased tissues or organs, in order to prevent failure using natural or tissue engineered materials, is the essence of regenerative medicine. Technologies can involve the use of stem cells, gene therapy, tissue engineering, and the use of artificial organs. To date, researchers have been recreating a range of tissues and organs in vitro, with varying degrees of success.

The tools to create scaffolds and structures are limited, and fine control over mechanical and environmental variables is complex. Also, there is a limited means of diagnosing at the molecular, cellular, and tissue levels what is actually happening within the constructs. In this chapter, the focus is on advancements in different areas of tissue engineering and regenerative medicine starting with the importance of scaffolds for tissue engineering. Further described, strategies for nerve regeneration both in the PNS and CNS and end in a brief review of sources of multipotent cells for tissue engineering, including remarks on the use of genetic techniques for in vitro differentiation.

In: Tissue Engineering
Editors: T. Ishikawa and M. Yamamoto
ISBN 978-1-62257-062-1

Chapter 1

LIVER REGENERATIVE MEDICINE: ADVANCES AND CHALLENGES

Dimitry A. Chistiakov*
Pirogov State Medical University, Russia

ABSTRACT

Liver transplantation is the standard of care for many end-stage liver diseases. But donor organs are scarce, and some people succumb to liver failure before a donor is found. Liver regenerative medicine is a special interdisciplinary field of medicine focused on the development of new therapies incorporating stem cells, gene therapy, and engineered tissues in order to repair or replace the damaged organ. In this review, we consider the emerging progress achieved in the hepatic regenerative medicine within the last decade. The review starts from the characterization of liver organogenesis, fetal and adult stem/progenitor cells. Then, applications of primary hepatocytes, embryonic and adult (mesenchymal, hematopoetic, and induced pluripotent) stem cells in cell therapy of liver diseases are considered. Current advances and challenges in producing mature hepatocytes from stem/progenitor cells are discussed. A section about hepatic tissue engineering includes consideration of synthetic and natural biomaterials in engineering scaffolds, strategies and achievements in the development of 3D bioactive matrices and 3D hepatocyte cultures, liver microengineering, generating bioartificial liver and prospects in fabrication of the bioengineered liver.

ABBREVIATIONS

ABC	ATP-binding cassette transporter
ASGPR1	asyaloglycoprotein receptor 1
BLSS	Bioartificial Liver Support System
BMP	bone morphogenetic protein

* Correspondence concerning this article should be addressed to: Dimitry A. Chistiakov, Department of Medical Nanobiotechnology, Pirogov State Medical University, Ulitsa Ostrovityanova 1, 117997 Moscow, Russia. Tel. +7(495) 434-13-01; fax +7(495) 315-05-01; E-mail: dimitry.chistiakov@lycos.com.

CD34	cluster of differentiation molecule 34
CK	cytokeratin
c-kit	tyrosine-protein kinase Kit (mast/stem cell growth factor receptor, protooncogen c-kit, CD117)
c-Myc	Myc protooncogen protein
CYP3A4	cytochrome P450, family 3, subfamily A, member 4
FGF	fibroblast growth factor
Foxa3	forkhead box A3 (hepatocyte nuclear factor 3-gamma)
dlk-1	delta-like kinase 1
ECM	extracellular matrix
EGF	epidermal growth factor
ELAD	extracorporeal liver assist device
EpCAM	epithelial cell adhesion molecule
ESC	embryonic stem cell
GATA-6	GATA-binding factor 6
G-CSF	granulocyte colony-stimulating factor
Hex	hematopoetically expressed homeobox factor
HGF	hepatocyte growth factor
HNF	hepatocyte nuclear factor
HSC	hematopoietic stem cell
IL-6	interleukin 6
iPSC	induced pluripotent stem cell
Ki67	antigen identified by monoclonal antibody Ki-67 (MKI67)
Klf4	Krüppel-like family of transcription factors, member 4
Liv2	fetal hepatoblast surface marker
MELS	Modular Extracorporeal Liver Support
MSC	mesenchymal stem cell
MiR-194	microRNA 194
Nanog	nanog homeobox (embryonic stem cell transcription factor)
NHSBT	National Health Service Blood and Transplant
Oct-4	octamer-binding transcription factor 4
OPTN	Organ Procurement and Transplantation Network
p19/Arf	ARF tumor suppressor (transcribed from an Alternate Reading Frame of the INK4a/ARF locus)
PF	polyurethane foam
PLLA	polyglycolate-polylactate (copolymer of polyglycolic and polylactic acids)
Prox-1	prospero-related homeobox factor 1
Rex-1	reduced expression protein 1 (zinc finger protein 42)
Sca-1	ataxin-1
Sox-2	SRY (sex determining region Y)-box 2
Tbx-3	T-box transcription factor 3
TGF beta	transforming growth factor beta
Thy1	THYmocyte differentiation antigen 1
TNF alpha	tumor necrosis factor alpha
WNT9a	Wingless-type MMTV integration site family, member 9A

INTRODUCTION

In 1963, Thomas E. Starzl, an American physician and surgeon, performed the first human liver transplantation in the University of Colorado Health Sciences Center [1]. Then, he founded the world first center of organ transplantation in Pittsburgh, PA, USA. In late 1980s', over 500 liver transplantations were annually performed in this center. In Europe, the first liver transplantation center (and second in the world) was organized and headed by Roy Y. Calne in Cambridge, UK in 1967 [2].

Advances in liver surgery, storage, and rapid organ delivery along with opening new liver transplantation services worldwide resulted in continuous increase in the number of transplantations. In 1997, around 8,000 liver transplantations were performed in the world. At present, this number exceeds 11,000. USA is the world largest contributor in this field achieving 6,300-6,650 liver transplantations per year in a period of 2005-2009 [3]. In Europe, about 4,000 donor livers are annually transplanted including 600-800 operations in Germany [4] and 600-670 transplantations in UK [5]. However, despite the constantly growing number of transplantations, the annual need in liver donation is covered by 30-50% only.

Another serious limitation is the high cost of liver transplantation. According to the United Network for Organ Sharing, average US cost estimates for liver transplantation during the first year are $523,400 [6], and these expenses stably rise. In addition, current strategies of liver transplantation consider the essential implementation of immune suppression to minimize the risk of organ rejection. Despite the obvious progress in the immunosuppressive techniques resulted in constantly increasing survival of transplant patients, the risk of developing acute infection and wound complications still persists [7-9].

Therefore, it is necessary to develop alternate strategies for liver replacement, regeneration, and repair, which are cost-effective, independent on the shortage of donor livers and involve using autologous liver cells for engraftment to avoid transplant rejection and post-transplantation complications. In this review, we will consider current trends and challenges in modern liver regenerative medicine focused on the generation of these advanced approaches.

LIVER EMBRYOGENESIS

To date, a substantial experimental and theoretical knowledge was achieved in liver regenerative medicine. Nevertheless, to successfully develop new efficient approaches for liver regeneration, in-depth knowledge about molecular and cellular mechanisms of liver embryogenesis and self-repair and, of particular, a key role of pluripotent stem cells and cell precursors in the repair of hepatic tissues is vitally essential.

During embryogenesis, totipotent embryonic stem cells (ESCs) of the blastocyst inner cell mass form three germ layers consisting of multipotent endodermal, mesodermal, and ectodermal cells. The embryonic liver develops from the lateral domains of the definitive endoderm in the ventral foregut [10] and from a small population of cells tracking down the ventral midline [11]. During foregut closure, the medial and lateral domains fuse together and the endoderm cells are specified to a hepatic fate. This tissue specification happens under well-coordinated action of inductive signals and genetic regulatory factors that are highly conserved among vertebrates. These developmental influences, which are critical for hepatic induction, include signals mediated by fibroblast growth factors (FGF) from the cardiac mesoderm and bone morphogenetic proteins (BMP) from the septum transversum mesenchyme [12].

The hepatic induction starts from the derepression and activation of Wnt-dependent signaling that is essential for differentiation of hepatic cell precursors into fetal liver cell called hepatoblasts [13, 14]. Like mature hepatocytes, hepatoblasts have a columnar shape. They migrate to the septum transversum mesenchyme to form the liver bud [15]. At the stage of budding, hepatic development is tightly regulated by a network of new transcription factors including hepatocyte nuclear factor HNF-6, T-box transcription factor Tbx-3, hematopoetically expressed homeobox factor Hex, prospero-related homeobox factor 1 Prox-1, and GATA-6 [16]. Mesenchymal cells secrete a range of cytokines and growth factors such as FGF, epidermal growth factor (EGF), hepatocyte growth factor (HGF), transforming growth factor (TGF) beta, tumor necrosis factor (TNF) alpha, and interleukin-6 (IL-6) supporting proliferation of hepatoblasts [17]. The presence of endothelial cells, which are independent of the blood supply and secrete specific signaling molecules, is also crucial for normal liver organogenesis. For example, observations of chicken liver embryogenesis showed that Wnt9a secreted by the hepatic sinusoids not only stimulates hepatoblast proliferation, but it also controls global liver morphologenesis [18].

Liver bud hepatoblasts are bipotential cells capable to differentiate to hepatocytes and biliary endothelial cells (cholangiocytes). Hepatocytes express specific markers such as albumin, transthyretin, and alpha-fetoprotein while cholangiocytes express cytokeratin CK-19 [19]. Both HGF and oncostatin M control differentiation of hepatoblasts to hepatocytes [20] whereas the Jagged-Notch signaling is essential for developing of a biliary epithelial phenotype [17]. On final stages of liver development, the percentage of bipotent cells gradually reduces resulting in irreversible formation of either unipotent hepatocytic or cholangiocytic lineage.

Strong knowledge about the developmental stages of liver organogenesis and sets of specific growth factors influencing proliferation and differentiation of liver cell progenitors at each of these stages has essential practical applications. For example, it is necessary to know sequential effects of different tissue-specific factors on hepatic cell precursors in order to increase efficiency and yield of functional hepatocytes from cultivating embryonic and adult stem cells. In one of the next sections of this review, we will consider current strategies and challenges of directional differentiation of human ESCs and other stem cells to mature hepatocytes and hepatocyte-like cells.

Fetal Liver Stem/Progenitor Cells

Fetal liver progenitor cells hold significant potential for future clinical use since they capable for self-renewal and bipotent differentiation to hepatocytes and cholangiocytes. Compared to ESCs and induced pluripotent stem cells (iPSCs) whose multi-step differentiation to hepatic cells is still challenging, fetal progenitor cells are already committed to a hepatic fate. Indeed, compared to ESCs, no substantial efforts are required to convert fetal progenitor cells to mature liver cells.

Liver progenitor cells of human fetus express several surface markers such as delta-like kinase dlk-1, E-cadherin, Thy1 and Liv2 [21, 22] thereby suggesting for the possibility of selective isolation of these cells via cell sorting. Human hepatic progenitors were already isolated from fetal livers using immunoselection against the surface marker, an epithelial cell adhesion molecule EpCAM [23]. EpCAM-positive cells were shown to differentiate into both hepatic and biliary cell lineages [24] and differentiate into the mature human liver tissue in adult rats with liver injury [25]. The EpCAM-positive progenitors are located in the ductal plate of the fetal liver. Interestingly, EpCAM-positive cells were also found in the adult human liver [26] and hepatocellular carcinoma [27] suggesting that this cell adhesion molecule is specific for both adult hepatic stem cells [28] and hepatic carcinoma stem cells [29].

In the fetal liver, side population cells are another potential stem cell population. These cells were reported to contribute to early stages of fetal liver development giving induction of the hematopoietic and epithelial lineages during 2nd semester [30]. These cells were first isolated due to their ability to accumulate fluorescent DNA-staining dye Hoechst 33342 via the transfer activity of ATP-binding cassette (ABC) transporters [31]. Fetal side population cells are suggested to belong to the mesenchymal-epithelial transitional lineage probably derived from the mesendoderm since they, additionally to bile ductal cells, are able to differentiate to various tissues of mesodermal origin such as bone, fat, cartilage and epithelial cells [32]. To date, side population cells are poorly characterized, and the location of these cells in the liver is still unclear. Side population cells positive for ABCG2 were found in hepatocellular carcinoma and hepatoma cell lines [33, 34]. These cells are considered as potential stem cancer cells contributing to invasiveness and chemoresistance of liver tumors [35]. The putative tumorigenic potential of fetal liver progenitors limits their utility for clinical use. Indeed, further careful examination of properties of human fetal stem/progenitor cells in tissue culture and animal models of liver injury is required before making a final decision about their suitability for clinical practice.

Adult Liver Stem/Precursor Cells

In the adult liver of mammals, four populations of candidate stem cells were detected including so called oval cells, small hepatocytes, epithelial liver cells and mesenchymal-like cells. Oval cells were found in rodents. Mass proliferation of hepatic oval cells is observed in a variety of liver injuries suggesting about a key role of this precursor cell population in liver repair [36]. The ability to produce bile duct markers (CK-7, CK-19) and hepatocyte markers (alpha-fetoprotein and albumin) suggests a precursor/product relationship between oval cells and hepatoblasts [37]. Activation of stem cell genes such as c-kit, CD34, dlk1, and others observed in oval cells suggests that these cells display stem cell-like properties [38, 39]. The finding that oval cells also express hematopoietic stem cell (HSC) genes (such as c-kit, CD34, Sca-1 and Thy-1) led to the suggestion that they originate from HSCs [40]. However, several subsequent studies in mice and rats showed that transdifferentiation of HCS into oval cells is a very rare event, which probably does not have physiological significance [41, 42]. Hepatic oval cells are likely to arise from non-parenchymal (ductal) cell lineage and represent a population of undifferentiated epithelial cells in the periportal areas capable to proliferate and transdifferentiate into hepatocytes thereby contributing to liver regeneration [43, 44].

Liver oval cells were referred to as facultative stem cells or as a reserve stem cell compartment, which are activated in liver injury to participate in organ repair [45]. Once isolated, cultivating oval cells are capable to maintain self-renewal potential for a long time [46]. Several attempts were reported to evaluate the regenerative potential of rodent oval cells in experimental models of liver injury, however, the rate of liver repopulation achieved by the engraftment of oval cells was low [47-49]. Using genetically modified murine oval cells tranduced with adeno-associated virus vector expressing human alpha1-antitrypsin resulted in significantly higher rates (up to 40-50%) of liver repopulation [50]. At present, oval cells are widely used as analogs of human liver adult/precursor cells for cell therapy in experimental animal models of inherited human liver disorders [51].

Small hepatocytes were first derived by Mitaka et al. [52] from the adult rat non-parenchymal liver. Compared to typical hepatocytes, these cells have a smaller size and flat shape. Small hepatocytes can proliferate and differentiate to mature hepatocytes in vitro and in vivo [53, 54]. These cells were considered as 'committed progenitor cells' capable to differentiate to mature hepatocytes when the liver repair is urgently needed [55]. Small hepatocytes were used for reconstitution of hepatic organoids that consist of mature polarized hepatocytes and develop networks of bile canaliculi thereby mimicking the microstructure of the natural organ [56]. The organoids possessing all major functions of the native liver were used in early experiments to reconstitute the ultrathin tissue architecture of the intact organ in hepatic tissue engineering [57] and develop prototypic hepatic microorgans for studying drug metabolism and properties [58].

Population of epithelial liver cells with hepatobiliary phenotype was recently found in the non-parenchymal liver of adult men [59] and then in mice [60]. These cells expressed common markers for hepatic and stem cells such as Oct-4, Rex-1, CD90, CD44 and CD29 but were negative for hematopoietic cell markers CD34 and CD117 [61, 62]. The cells also produced not only immature (alpha-fetoprotein) but also differentiated hepatocyte (albumin and CK-18) and biliary markers (CK-7 and 19) [60]. However, epithelial liver cells were shown to lack the ability to express markers of mature hepatocytes such as cytochromes P450 and tyrosine aminotransferase. Epithelial cells hold capacity to differentiate towards hepatocyte-like cells exhibiting an intermediate hepatobiliary and mature/immature phenoltype [61]. Epithelial cells were considered to represent a resident progenitor population of the adult human liver, even in the absence of hepatic failure [61].

Mesenchymal-like progenitor cells isolated from the human liver expressed both hepatic and mesenchymal markers among which albumin, CYP3A4, vimentin, and alpha-smooth muscle actin [63]. Once transplanted into the immunodeficient mice, these cells differentiated into hepatocyte-like cells expressing hepatocyte-specific markers such as human albumin, prealbumin, and alpha-fetoprotein [63]. Mesenchymal-like progenitor cells expressing stem cell markers Thy-1, CD34, and Ki67 were also isolated from the liver of young and adult rats [64]. This cell population was able for in vitro differentiation to early cholangiocytes positive for CK19 and for differentiation to hepatocytes expressing CK18. However, upon changing growth conditions, mesenchymal-like could also produce non-liver cells such as adipocytes, astrocytes, neuroblasts, and osteoblasts [64].

For adult human progenitor cells isolated only several years ago, further studies of their properties, in vitro culturing conditions, and regenerative capacity are needed. Since these cell populations share many cell markers and properties, it is necessary to develop strict criteria that allow distinguishing, for example, between human mesenchymal-like cells and human epithelial liver cells. It would be interesting to know whether mesenchymal-like liver cells possess immunomodulatory properties like mesenchymal stem cells (MSCs), a feature that definitely has a big promise in liver cell therapy. Recently, Meng et al. [65] reported high levels of microRNA miR-194 in human liver epithelial cells preventing epithelial-mesenchymal transition of these cells and metastasis of liver tumors. This observation has a great potential for developing new targeted therapies against liver cancers.

Generation of Hepatocytes from Embryonic Stem Cells

To date, the use of human ESCs in liver regenerative medicine is strongly restricted by ethical barriers. In addition, too higher regenerative potential, which is difficult to control, and putative tumorigenicity also limit broad clinical application of ESCs. Indeed, the most efficient application of human ESCs is to serve as a visually unlimited source for obtaining mature hepatocytes and hepatocyte-like cells that could be then used for liver cell therapy, tissue engineering and other purposes.

At present, generation of hepatocyte-like cells from cultivated ESCs appear to represent the most well studied model of cell differentiation to mature hepatocytes. Within the last decade, a great variety of techniques of the generation of hepatocytes from ESCs was developed. Current strategies of obtaining ESCs-derived hepatocytes utilize spontaneous differentiation of ESCs to mature hepatocytes through formation of embryoid bodies in bioreactors and directional differentiation by sequential stimulation with different tissue-specific growth factors and cytokins.

In suspension, culturing human and mouse ESCs were shown to spontaneously aggregate to embryoid bodies consisting of three germ layers: ectoderm, mesoderm, and endoderm [66, 67]. Although forming embryoid bodies reproduce the natural microenvironment essential for further differentiation of ESCs to liver-specific cell lineages, the differentiation yields cell populations of mixed origin arisen from all germ layers [68]. Heterogeneity of cell populations resulted from spontaneous differentiation of embryoid bodies greatly hampers further isolation of hepatocytes.

Since hepatocytes are originated from the definitive endoderm, it is necessary to enrich this fraction in embryoid bodies [69]. The enrichment could be achieved by treatment of cultivating ESCs with a mixture of early embryonic differentiation factors such as activin A, Wnt3, fibroblast growth factor FGF2 (or FGF4), and bone morphogenetic protein BMP2 (or BMP4) [70-75]. FGF2/4 and BMP2/4 support further preferential differentiation of the definitive endoderm towards hepatic cell lineages resulting in increase in yield of mature hepatocytes up to 70% [73].

A strategy of the directional differentiation of ESCs to functional hepatocytes suggests the sequential supplement of the growth medium with cytokines and growth factors involved in the embryonic development of the mammalian liver. The growth factors typically used in the directional differentiation include a family of FGFs (stimulate generation of bipotent hepatoblasts from cultivated ESCs and their maturation to hepatocytes), HGF (supports formation of hepatocytes from hepatoblasts), follistatin (essential for differentiation of hepatoblasts to ductal cells), and dexamethasone (glucocorticoid hormone inducing production of proteins specific for adult hepatocytes) [76-78]. Implementation of this strategy in cultivating ESCs provides avoiding aggregation of cells to embryoid bodies [79] and increasing ratio of conversion of ESCs to mature human hepatocytes up to 80% [77] or even up to 90% [74].

On the final step of obtaining ESCs-derived hepatocytes, mature hepatocytes should be separated from contaminant cells representing a heterogeneous population of hepatic precursors and other cell types at intermediate stages of differentiation and maturation. Genetic modification of ESCs with transgenic constructs driven by a hepatocyte-specific promoter (for example, alpha-fetoprotein promoter) provides possibility for highly selective isolation of hepatocytes using cell sorting [80]. However, clinical applications of genetically modified hepatocytes are restricted by current biosafety and bioethical concerns.

Nevertheless, hepatocytes have a specific surface marker, asyaloglycoprotein receptor ASGPR1, responsible for binding of partially deglycosylated serum proteins and immunoglobulin A2 for further transferring and degradation of these proteins in lysosomes [81]. Targeting ASGPR1 with fluorescent-labeled or magnet-coated monoclonal antibodies allows subsequent higher efficient isolation of ESC-derived ASGPR1-positive hepatocytes [82].

In the future, further efforts should performed towards improvement of experimental protocols of the directional differentiation of ESCs to hepatocytes to achieve almost 100% or even complete ratio of hepatic-specific conversion of ESCs that will reduce costs and allow to avoid using expensive procedures of isolation and enrichment of hepatocyte fraction with help of flow cytometry. Recent implementation of 3D bioengineered scaffolds results in markedly accelerated differentiation and maturation of ESC-derived hepatocytes with improved metabolic and functional properties. This evidence is based on the results of multiple studies showed a variety of benefits of 3D cultures of ESCs growing in collagen scaffolds [83], polylactate-polyglycolide polymeric matrices [84], alginate microbeads [80], and other engineered scaffolds compared to ordinary 2D monolayer cultures. Stimulatory effects of bioengineered scaffolds on the maturation rate and functional characteristics of hepatocytes can be explained by beneficial properties of the microscaffold architecture that reproduces the native structure and microenvironment of the liver parenchyma.

It is necessary to develop 3D cultures of ESCs with new advantageous properties in order to increase the differentiation rate, viability and functional features of hepatocytes. For example, this includes generation of new 'smart' biomaterials capable to respond to biological environment stimulation and avoid non-specific protein adhesion. Such 'smart' materials responding to specific cellular signals include hydrogels containing matrix metalloproteinase degradable sites and integrin binding sites [85] and scaffolds composed of self-assembling peptides [86]. 3D matrices consisting of self-assembling peptides were shown to be able to support directed differentiation of ESCs to hepatic-specific cell lineages [87].

MESENCHYMAL STEM CELLS IN LIVER CELL THERAPY AND AS A SOURCE FOR PRODUCTION OF HEPATOCYTES

Non-chematopoietic MSCs could be derived from various adult tissues including non-stromal bone marrow, adipose tissue, amniotic fluid, and dental pulp. They hold a great proliferation potential and a multipotent capacity to differentiate to a variety of cell types including hepatic cells. The ability of MSCs to differentiate to hepatocytes was first shown by Lee et al. [88]. In undifferentiated MSCs derived from the adipose tissue, expression of several hepatic markers (alpha-fetoprotein, CK-14, CK-19, HNF4) was found therefore suggesting for the presence of precursors of liver cell lineages [89].

By treatment with a cocktail of growth factors involved in liver organogenesis, researchers were able produce functional hepatocyte-like cells from MSCs of various origin in vitro. For example, to generate hepatocytes from MSCs of mixed origin (from umbilical cord blood and bone marrow), stem cells were cultivated in the presence of FGF and basic FGF that were replaced by a mixture of HGF and FGF [88]. To obtain hepatocytes from fat-derived MSCs, those were treated with HGF and oncostatin M [90]. Formation of hepatocytes from bone marrow-derived MSCs was induced either with endothelial growth factor EGF [91] or with sequential stimulation with HGF+FGF4 and oncostatin M+ dexamethasone [92]. Once engrafted to immunodeficient animals with hepatic failure, MSCs-derived human hepatocytes were able to repopulate liver cells displaying functional properties comparable with those of primary human hepatocytes [90, 91, 93-95].

In liver cell therapy, MSCs are tested to treat liver cirrhosis. In phase I and II clinical trials, injection of autologous bone marrow MSCs into into peripheral or the portal vein resulted in significant improvement in liver function as reflected by restoration of normal levels of hepatic production of albumin, creatinine, and bilirubin and absence of complications and acute side effects [96, 97]. Immunomodulatory activity of MSCs also has a great promise in the treatment of acute graft-versus-host disease. For example, in children with intestine transplants experienced acute, steroid-resistant graft-versus-host disease with the involvement of liver and skin, regular injection of autologous bone-marrow MSCs led to favorable results in 75% patients [98]. Overall, MSCs-based treatment of liver cirrhosis and other chronic hepatic diseases holds a huge therapeutic potential. Despite highly promising preliminary results, before a wide clinical use, the regenerative capacity of autologous human MSCs has to be further extensively evaluated in ongoing and future clinical trials.

LIVER CELL THERAPY WITH HEMATOPOIETIC STEM CELLS

HSCs can be isolated from the adult bone marrow stroma, peripheral and umbilical cord blood. These cells represent a heterogeneous cell population carrying specific surface markers ($CD150^+$ $CD48^-$ $CD244^-$ lin^-). HSCs are primarily responsible for giving rise to all blood cell types of lymphoid and myeloid lineages but also capable to differentiate to non-hematopoietic cell types. The capacity of murine and human MSCs to produce hepatocyte-like cells able to regenerate liver tissues was shown in multiple experiments involving animal models of liver failure [92, 99-102].

Compared to non-hematopoietic MSCs, clinical trials with the involvement of human HSCs started earlier on. For treatment of cirrhotic patients, autologous CD34+ HSCs are injected to blood, with subsequent mobilization by granulocyte colony-stimulating factor G-CSF. In most clinical trials, beneficial results were achieved, with prolonged (up to one year) improvement of liver function after treatment with HSCs [103-107]. However, in certain circumstances, an unfavorable ability of HSCs to preferentially differentiate to myofibroblast-like cells (not to hepatocytes) was observed in a murine model of liver injury [108] and in the liver of cirrhotic patients [109] thus contributing to hepatic fibrosis. Therefore, the pro-fibrogenic potential of human MSC transplant should be not under-evaluated in further clinical trials.

Developmental Potential of Induced Pluripotent Stem Cells

The generation of human iPSCs through reprogramming of adult somatic cells with few transcription factors (Oct 3/4, Klf4, Nanog, Sox2, c-Myc) responsible for maintaining pluropotency of stem cells opens new horizons in regenerative medicine. To date, iPSCs were produced from primary hepatocytes from mice [110] and humans [111]. Accordingly, hepatocyte-like cells have been produced from iPSCs [112-115].

Obviously, iPSCs represent an unlimited source for producing mature hepatocytes. However, due to retaining an 'epigenetic memory' of their cell type of origin, low-passage iPSCs of mesodermal origin should spontaneously differentiate to embryoid bodies preferentially consisted of mesodermal cells [116], a property that is not beneficial for producing hepatocytes arisen from the definitive endoderm. Therefore, it would be better to produce iPSCs directly from hepatocytes and other endodermal cells in order to increase the homogeneity of terminally differentiated hepatocyte cultures generated from these iPSCs. In theory, treatment with hepatic growth factors should also stimulate preferential differentiation of iPSCs of mesodermal origin to liver cells. However, attempts of Sancho-Bru et al. [117] obtain hepatocyte-like cells from murine iPSCs by directed differentiation resulted in producing a mixture of hepatocytes and mesodermal cells displaying some characteristics of liver sinusoidal endothelium and stellate cells.

Ethical limitations applied to human ESCs cannot be extended to human iPSCs since production of human iPSCs does not involve the destruction of a human embryo [118].

In the nearest future, the most promising field of application of iPSCs appears to be the generation of iPSCs-derived hepatocytes. To date, substantial efforts were performed to increase the efficiency of reprogramming and yield of iPSCs from somatic cells via inhibiting negative cell cycle regulators [119, 120]. To increase the biosafety of produced iPSCs, retroviral vectors for ectopic expression of pluripotency factors have been replaced with non-viral vehicles [121, 122] and self-inactivating constructs [123], and potential protooncogens Klf4 and c-Myc were withdrawn from new reprogramming protocols to produce iPSCs with as few as even one pluripotency factor [124]. Recently, functional hepatocyte-like cells have been successfully produced through reprogramming mouse adult fibroblasts by transduction of Gata-4, Hnf1alpha and Foxa3, and inactivation of the negative cell regulator p19/Arf [125]. Thus, iPSC technology seems to be a viable method for generating hepatocytes, without the controversy surrounding use of embryonic cells. However, similar to the case with ESCs, much additional work is needed to define the methodologies necessary to achieve liver-specific differentiation.

The next promising field for using iPSCs in hepatology is studying regenerative capacity of iPSC-derived hepatocytes in animal models of various liver pathologies and development of cell-based panels for modeling hepatic diseases. Of particular, iPSCs will be useful for modeling of inherited liver syndromes and correction (gene therapy) of genetic defects causing these syndromes [126].

Another perspective field for using iPSC-derived hepatocyte-based cell panels is a high throughput screening of potential drugs and massive drug testing for toxic and adverse effects. Human hepatocarcinoma cell lines and animal hepatocytes widely used in pharmacokinetics/ pharmacodynamics studies cannot fully reproduce biochemical and physiological effects of drugs to natural human hepatocytes. Therefore, application of iPSC-derived human hepatocytes whose properties are close to those of native adult hepatocytes should significantly improve the efficiency of drug screening and evaluation and decrease drug toxicity and risk of drug-induced side effects.

NON-STEM CELLS (PRIMARY HEPATOCYTES)

Hepatocytes are unique among differentiated parenchymal cells because they retain a stem cell-like ability to proliferate [45]. This property underlies the remarkable capacity of liver to regenerate following acute injuries that diminish hepatic mass. Hepatocyte proliferation is also an important adaptive response in chronic liver diseases as hepatocytes destroyed by inflammation and other injuries are replaced to maintain adequate liver function.

The ability of hepatocytes to proliferate in stress conditions was widely used in pioneer works in liver cell therapy. In 1992, the first transplantation of allogenic human hepatocytes with normal gene encoding the low density lipoprotein receptor was performed to patients with familial hyperholesterolemia [127]. Till 2006, a total of 78 human clinical hepatocyte transplant experiences were reported [128]. Overall, the results of hepatocyte-based cell therapy were inconsistent. For example, treatment of hepatic disorders caused by mutations in urea cycle enzymes (ornithine transcarbamylase deficiency, etc.) with normal human hepatocytes led only to the temporary improvement in liver function (e.g., restoration of urea synthesis) followed by the relapse of disease symptoms due to the transplant rejection [129, 130] or due to another reason [131], which finally progressed to liver failure and need in organ transplantation [132]. For therapy of non-genetic liver disorders (cirrhosis, acute liver inflammation), conflicting results were also obtained, with either favorable [133-135] or fatal outcomes [133, 136].

The efficiency of engraftment of hepatocytes is markedly reduced by immunosuppression before the transplantation. For example, rapamycin significantly decreases integration and proliferation of hepatocyte transplant in the recipient tissue [137]. Indeed, to increase survival of the hepatocyte graft, it would be better to transplant autologous cells in order to avoid an immunosuppressive therapy.

Using genetically modified hepatocytes transduced with a lentivirus-based vehicle carrying the normal transgene is considered as an alternative strategy for treatment of inherited liver disorders (if approved for clnical use) [138, 139]. Genetically modified hepatocytes showed advanced metabolic and regenerative properties in pre-clinical trials exhibiting high efficient (up to 90%) integration to the parenchymal liver and stable expression of a normal gene copy in animal models of human inherited liver diseases [140, 141]. However, to proceed to clinical testing of these cells, it is necessary to resolve current biosafety requirements limiting or even prohibiting clinical application of any bioproduct of viral origin.

Although many advanced techniques of generation of hepatocytes from embryonic and adult stem cells were developed, primary hepatocytes maintain their utility in pharmacology and regenerative medicine including hepatic tissue engineering and constructing bioartificial liver [142, 143].

Extracorporeal Bioartificial Liver Support Systems (Bioartificial Liver)

Bioartificial liver is an extracorporeal device filling with primary hepatocytes for temporary support of liver function in a patient waiting for the liver transplantation. Ideally, this supporting system should fulfill all hepatic functions and be biosafe. Unfortunately, all cell sources so far used for bioartificial liver support do not completely fulfill the above criteria [144].

First attempts to use animal liver tissues to support patients with the end-stage liver disease were performed in 1950s' [145]. The mass application of this approach began in 1980s', when donor livers found to be unsuitable for transplantation due to steatosis, cirrhosis, or traumatic injury, started to be used for temporary liver support of a recipient patient [146]. Despite a highly promising start, using extracorporeal liver support systems revealed several serious problems including quick loss of functional activity and limited ex vivo viability of primary hepatocytes [147]. Bioartificial liver device AMC developed in Academic Medical Center (Amsterdam) and based on using pig hepatocytes showed favorable results in a phase I clinical trial including the complete recovery of one patient and no need in further liver transplant [148]. However, Fruhauf et al. [149] isolated endogenous reovirus from primary porcine hepatocytes used in the AMC machine capable to infect culturing primary human hepatocytes. The finding resulted in prohibiting use of these devices in several European countries [150].

To date, several extracorporeal liver support systems were developed including Hepta Assist, Modular Extracorporeal Liver Support (MELS), Bioartificial Liver Support System, and Radial Flow Bioreactor. Most of these devices involved now in phase 1 clinical trials are charged with primary human hepatocytes in order to prevent transmission of zoonotic infections [151]. To support prolonged ex-vivo survival and functionality of liver cells inside these systems, a variety of technical advances were implemented. In the MELS system, albumin dialysis is used for the removal of albumin-bound toxins, reducing the biochemical burden of the liver cells and replacing the bile excretion of hepatocytes in the bioreactor [152]. In Radial Flow Bioreactor, medium flow is directed from the perimeter to the centre or vice versa, thus providing advantageous concentration gradients throughout the bioreactor. Cells within the system are seeded on microcarriers enabling a 3D high-density culture [153].

In the bioartificial liver support system (ELAD - extracorporeal liver assist device) developed by Sussman et al. [154], immortalized human hepatocyte cell line C3A derived from the hepatoblastoma cell line is used. As expected, the use of immortalized human hepatocytes should resolve the problem of the regular replacement of non-functional hepatocytes with fresh cells. The ELAD system was evaluated, mostly with beneficial results, in several small clinical trials [155]. Several other immortalized hepatic cell lines are now under investigation to be used in bioartifical liver devices [156, 157].

In the future, it is necessary to increase the efficiency of a bioartificial liver by enhancing blood flow rate through a machine (in order to achieve more complete blood clearance). In current liver support devices, the flow reaches maximum 300 mL/min whereas in a normal human liver over 1.5 L of blood passes per minute. This advance is now realized in the Bioartificial Liver Support System (BLSS) that, in opposite to other devices using blood plasma perfusion, utilizes the whole blood perfusion. After preclinical testing in dogs [158], BLSS was recommended to proceed to clinical testing [159].

Another challenge considers oxygenation/gas supply and using matrices to provide 3D microenvironment for hepatocytes since it is crucial for survival and metabolic activity of liver cells. Most bioartificial liver devices use the direct growth medium oxygenation. For example, in the Radial Flow Bioreactor, medium is recirculated through the bioreactor by a roller pump and oxygenation is achieved by bubbling 95% oxygen/5% carbon dioxide through the medium reservoir [160]. In most bioartificial liver support divices, hepatocytes are immobilized in scaffolds in the form of hollow fibres. Within scaffolds, hepatocytes reform cell aggregates, in the middle of which cells could be exposed to hypoxia and therefore to enhanced loss of functionality and death. To overcome this obstacle, it is possible, for example, to encapsulate hepatocytes in beads or microcapsules consisting of a hydrophilic gel such as alginate gel [161]. Within a hydrophilic gel, hepatic cell aggregates were shown to have enough supply with nutrients and oxygen to survive for a long time [162]. Furthermore, applying rotary systems mimicking microgravity accelerate maturation and functional polarization of hepatocytes as happen in the native liver [163]. A new prototype rotary system-based bioartificial liver charged with hepatic progenitor cells (small human hepatocytes), which is now under preclinical evaluation, showed advantageous properties in supporting a stable viability (over 90%) of these cells against treatment with alcohol, benzodiazepines, and other toxic drugs [164-166].

A ROLE OF THE EXTRACELLULAR MATRIX IN LIVER REPAIR AND HEPATOCYTE FUNCTIONALITY

To reach the maximum functionality, stem and non-stem cells require a special microenvironment called cell niche. A niche represents a complex microcommunity of cells, extracellular matrix (ECM), and a network of signaling molecules released by cells for maintaining cell integrity within a niche. ECM plays a key role in supporting cell integrity, viability, growth, and function [167]. Alterations in ECM influence differentiation of human embryonic and adult stem cells [168]. Cell niches have a 3D architecture biochemical and topological changes of which deeply affect cell differentiation and maturation [169].

Collagens, elastin, proteoglycan, and structural glycoproteins are the major ECM components. Cells are integrated into a 3D framework of a high complexity constituted by these proteins. The complex between cells and ECM is called a 3D matrix, which supports cell-cell contacts and adhesion, delivery of nutrients, removal of waste products, and local networks of extracellular signaling. Integrins, a family of surface transmembrane hetoro-dimermolecules, are responsible for ECM-cell contacts [170].

Liver ECM is well studied due to its crucial role in maintaining hepatic homeostasis and liver cirrhosis pathogenesis [171]. Mature hepatocytes are characterized with unique cell polarity as reflected by intracellular compartmentalization of cytoskeleton and organells and subdivision of a plasmatic membrane to three morphologically distinct sections: basolateral, canalicular, and lateral [172]. The basal surface of hepatocytes is responsible for transfer of low-molecular compounds and metabolites while bile acids and cell wastes are secreted through the apical domain. Therefore, the possibility to modulate polarity and hierarchical organization of hepatocytes in the liver would be critical in constructing in vitro liver assist systems capable to successfully replace hepatic functions.

The ECM derived from the liver was shown to be able to support functions and properties of hepatic cell co-cultures [173]. In constructing new biomaterials and scaffolds for a bioartificial liver, researchers and tissue engineers try to maximally reproduce physical and chemical characteristics of the natural liver 3D ECM in order to extend survival and functionality of hepatocytes [174]. There are many examples showing advantageous properties of hepatic cells growing in 3D scaffolds compared to ordinary monolayer cultures [175-177]. The limitations of existing 2D tissue culture techniques have necessitated the creation of 3D microenvironments incorporating materials, implants, and biophysical stimuli optimized for modulation of cell functions.

Bioengineering 3D Scaffolds from Natural Biomaterials

The modern tissue engineering is focused on the development of 3D advantageous scaffolds that mimic unique properties of natural ECM as much as possible. In liver tissue engineering, modified structural components of the natural ECM are widely used in constructing bioactive scaffolds. The bioactive scaffolds are capable to stimulate cell growth and differentiation. Collagen is the major structural protein in cartilage, bone, and skin tissues that supports cell adhesion, attachment, proliferation, growth, differentiation, and migration [178]. Similarly, collagen-based scaffolds were capable to efficiently promote in vitro differentiation of human ESCs-derived embryoid bodies to hepatocyte-like cells phenotypically and functionally close to primary human hepatocytes [83, 179].

Hyaluronic acid is another key component of ECM that is involved in the regulation of cell expansion and proliferation and plays a primary role in mediating cell-cell contacts and cell behavior (e.g., migration). Fetal hepatoblasts and adult hepatocytes express on their surface the CD44 molecule, a receptor for hyaluronic acid [180]. The presence of CD44 provides the possibility to create hydrogels consisting of hyaluronic acid and its chemically modified derivatives with improved adhesion properties for hepatocytes and hepatoblasts and capable to support viability and functionality of hepatic cells during four weeks [180].

In generating bioengineered matrices, a variety of natural biomaterials including chitin, chitosan, silk, alginate, matrigel, and sponge, were applied. For example, silk fibroin-based microfluidic devices demonstrated beneficial properties in supporting proliferation and function of HepG2 cells [181]. Chitosan is a partially deacetylated hitin, which is composed by glucosamine and N-acetylglucosamine. A matrix engineered from chitosan and albumin supported the attachment of porcine hepatocytes that then developed hepatic organoids and tissue fragments [182]. Culturing primary hepatocytes in the scaffold containing chitosan nanofibers with incorporated glucose residues led to elongated metabolic activity of cell aggregates formed by hepatocytes [183]. Generation of a mixed matrix from chitosan and matigel (a natural gel) enriched with heparin molecules resulted in engineering a bioactive scaffold containing a depot of basic FGF, a growth factor essential for differentiation of hepatic cell lineages [184].

Using natural hydrogels such as alginate and matrigel significantly improves the biocompatibility of artificial matrices and enhances their seeding with cells. The Matrigel is composed by proteins (laminin, collagen type IV, heparansulfate proteoglycan, etc.) extracted from the basal membrane of murine chondrosarcoma [185]. Hepatocytes growing in matrigel aggregate to multicellular clusters consisted of cells whose shape is less polygonal than in natural hepatocytes [185]. However, a precise composition of the matrigel is unknown that indeed hampers strict experimental control in creating bioengineered matrigel-based matrices with suggested properties. Accorging to new standards of the good manufacturing practice, tumor and xenogenic origin of the matrigel limits its broad application in tissue engineering.

Alginate hydrogel is a polysaccharide containing alpha-D-guluronic and beta-D-mannuronic acids. Alginate is a product derived from algae and capable to form a gel in the presence of bivalent cations such as Ca^{2+}, Ba^{2+}, Sr^{2+} [186]. Alginate is widely used in biology, medicine, and pharmacology [187]. Treated with hepatic growth factors, embryoid bodies culturing in alginate microcapsules were found to successfully differentiate to both types of adult liver cells [188]. Cultivation in alginate microbeads significantly increased growth of human hepatoblasts and their maturation to hepatocytes [189]. Human hepatocyte cell line C3A grown in alginate matrices formed multicellular spheroids enzymatic activity of cytochromes P450 in which was much greater than in hepatocyte monolayer cultures [175].

Thus, use of natural polymers in 3D systems mimics in vivo histoarchitecture and provides easy manipulation of cells and tissues as well as promoting cell-matrix interactions.

These provide a microenvironment conducive to normal progenitor cell kinetics and enhance cell differentiation. The major disadvantages of using natural materials are the limited control over physiochemical properties that they offer, as well as immunogenicity, degradability, lack of reproducibility, and inconsistency in mechanical properties.

SYNTHETIC POLYMERS IN LIVER TISSUE ENGINEERING

Compared to natural polymers, synthetic biomaterials possess a wide range of properties that could be easily controlled and changed. A variety of synthetic polymers with advanced biocompatibility and biodegradability could be also developed. Biodegradable polymer scaffolds provide frameworks for tissue regeneration and cell transplantation and can be made to degrade over time. To date, biodegradable polymers commonly used in regenerative medicine include polylactic acid, poly-L-lactic acid, polyglycolic acid, polyanhydrides, polyfumarates, polyorthoesters, polycaprolactones, and polycarbonates [176].

Polylactic and polyglycolic acids and their copolymers (polyglycolate-polylactate, PLLA) as well as copolymers of L- and D-isomers of lactic acid belong to poly (alpha-hydroxy) acids [191]. In tissue engineering, PLLA were applied to construct 3D scaffolds, which then were successfully used for differentiation of fetal hepatoblasts to mature hepatocyes [192-194]. The main limitations of PLLA are chemical instability, surface erosion, and hydrophobicity [195]. The chemical instability of poly (alpha-hydroxy) acids results in formation of many hydrolysis products capable to induce the inflammatory response. The chemical modification of polymers, incorporation of proteins and special bioactive domains increases the biocompatibility of bioengineered matrices and improves adhesion scaffold properties stimulating cell attachment and migration and thereby facilitating liver tissue repair [196].

Polyurethanes also hold advanced properties for generating 3D hepatocytes cultures. To grow hepatocytes and hepatocyte-like in bioreactors, polyurethane foam (PF) is used. In foam pores, hepatocytes form multicellular spheroids with improved functional characteristics [197, 198]. PF was suggested to be used as supporting scaffold for hepatocytes in bioartificial liver [199]. The PF-based scaffolds were shown to support growth, proliferation and differentiation of hepatocyte precursors and adult hepatocytes [200, 201]. A bioactive collagen gel scaffold filled with PF and immobilized heparin/vascular endothelial growth factor capable to enhance growth of three types of hepatocytes (single hepatocytes, hepatocyte spheroids, and hepatoblasts) was suggested as a suitable biodegradable matrix to support integration and induce angiogenesis within the hepatocyte transplant in a hepatotectomized animal model [202].

Conclusion: Towards the Development of a Bioengineered Liver

A major roadblock to successful organ bioengineering is the need for a functional vascular network within the engineered tissue. Hepatocytes growing as cell aggregates or in high dense cultures experienced acute hypoxia, since diffusion within microspheres is limited with several hundred microns [203]. Chronic hypoxia is one of the major factors causing death of primary hepatocytes in bioartificial liver support devices. Enhanced oxygen availability elongates survival of hepatocytes in the bioartificial liver [204]. As mentioned above, multiple efforts were performed to develop matrices with increased oxygen performance such as alginate and polydimethylsiloxane hydrogels and scaffolds containing synthetic oxygen donors (perfluorocarbons, calcium dioxide, etc.) or immobilized hemoglobin [205]. However, these engineered scaffolds cannot completely replace capillary functions in the liver parenchyma.

Interactions between different types of cells, particularly between hepatocytes and endothelial cells, are essential for normal functioning of the liver parenchyma [206]. Hepatocyte clusters were shown to spontaneously interact with hepatic stellate cells, which migrate into the clusters and produce ECM. Then, endothelial and ductal cells form microvessels and bile ducts thereby restoring the structure of the normal liver [207].

Furthermore, coculturing hepatocytes and biliary cells have beneficial effects to both hepatic cell lineages [208, 209]. These findings attracted investigators to start developing 3D heterologous cell systems, a prototype of the bioengineered liver.

Cocultures of murine hepatocytes with non-parenchymal (ductal, stellate, and endothelial) hepatic human cells in matrigel were able to spontaneously develop cellular prototypes of bill ducts and capillaries and to form microsinusoid structures producing albumin and cytochrome P450 [210]. Hepatocytes and endothelial cells growing in 3D scaffolds produced multilayer fragments of the liver-like tissues composed by hepatocyte clusters and capillary-like microtubes of endothelial cells. These engineered structures were viable for 3 weeks [211]. Using a biodegradable apatite-fiber scaffold, Saito et al. [212] reported the generation of the liver organoid reconstructed by mouse-immortalized hepatocytes and nonparenchymal cells (sinusoidal endothelial cells and hepatic stellate cells) in a radial-flow bioreactor. After transplantation, the organoid that had a liver-like structure with high-density layers of hepatocytes and vessel-like structures successfully survived in the omentum and kidney of recipient mice.

Another way for generation of a prototypic bioengineered liver is using natural scaffolds produced from a decellularized whole liver. Careful decellularization with detergent perfusion results in preserving the extracellular matrix components and the intact vascular network [213]. The advantage of decellularized native matrix is that it retains, to an extent, the original composition, and the microstructure hence enhancing cell attachment and reorganization. Fetal human hepatocytes and endothelial cells were shown to successfully recolonize decellularized natural liver scaffolds produced from different species and differentate to adult cells expressing typical endothelial, hepatic, and biliary epithelial markers [214]. The recellularized liver grafts can be transplanted into rats supporting hepatocyte survival and function with minimal ischemic damage [215]. The organoids seeded with human cells could be useful as organ assist systems for temporary replacement of liver functions in patients needed in the liver transplantation [210].

Overall, the modern liver regenerative medicine stays on initial step of the way towards the development of bioengineered liver. Much more experimental work should be done to extend achievements in the fabrication of liver microorgans to the creation of hepatic organoids and then proceed to the construction of a whole bioenginnered organ. The development of liver organoids based on natural decellularized scaffolds may greatly help in enhancing progress towards generating the bioengineered liver since technique for obtaining natural decellularized matrices is well developed for a variety of organs [216] and has a successful clinical application [217].

REFERENCES

[1] Starzl, T.E., Marchioro, T.L., von Kaaulla, K.N., Hermann, G., Btittain, R.S., & Waddell, W.R. (1963). Homotransplantation of the liver in humans, *Surgery, Gynecogy & Obstetrics*, 117, 659-676.

[2] Calne, R.Y., & Williams R. (1968). Liver transplantation in man. I. Observations on technique and organization in five cases, *British Medical Journal*, *4*, 535-540.

[3] Organ Procurement and Transplantation Network (2009). Available at: http://optn.transplant.hrsa.gov/latestData/rptData.asp

[4] Eurotransplant International Foundation (2010). Annual Report 2010. Available at http://www.eurotransplant.org/cms/index.php?page=annual_reports.

[5] National Health Service Blood and Transplant) (2010). Transplant Activity Report 2009-2010. Available at http://www.uktransplant. org.uk/ukt/statistics/transplant_activity_report/ current_activity_reports/ ukt/activity_report_2009_10.pdf.

[6] Milliman Research Report (2008). 2008 US organ and tissue transplant cost estimates and discussion. Milliman, Inc. April 2008. Available at: http://publications.milliman.com/research/health-rr/pdfs/2008-us-organ-tisse-RR4-1-08.pdf.

[7] Mehrabi, A., Fonouni, H., Wente, M., Sadeghi, M., Eisenbach, C., Encke, J., Schmied, B.M. Libicher, M., Zeier, M., Weitz, J., Büchler, M.W., & Schmidt, J. (2006). Wound complications following kidney and liver transplantation, *Clinical Transplantation, 20*, 97-110.

[8] Watt, K.D., Burak, K., Deschênes, M., Lilly, L., Marleau, D., Marotta, P., Mason, A., Peltekian, K.M., Renner, E.L., & Yoshida, E.M.; Canadian Transplant Hepatology Outcomes Research Network (2006). Recurrent hepatitis C post-transplantation: where are we now and where do we go from here? A report from the Canadian transplant hepatology workshop, *Canadian Journal of Gastroenterology*, *20*, 725-734.

[9] da Cunha-Bang, C., Sørensen, S.S., Iversen, M., Sengeløv, H., Hillingsø, J.G., Rasmussen, A., & Lemaigre, F.P. (2009). Mechanisms of liver development: concepts for understanding liver disorders and design of novel therapies. *Gastroenterology*, *137*, 62–79.

[10] Chalmers, A.D., & Slack, J.M.W. (2000). The Xenopus tadpole gut: fate maps and morphogenetic movements, *Development*, *127*, 381–392.

[11] Tremblay, K.D., & Zaret, K.S. (2005). Distinct populations of endoderm cells converge to generate the embryonic liver bud and ventral foregut tissues, *Developmental Biology*, *280*, 87-99.

[12] Shin, D., Shin, C.H., Tucker, J., Ober, E.A., Rentzsch, F., Poss, K.D., Hammerschmidt, M., Mullins, M.C., & Stainier, D.Y. (2007). Bmp and Fgf signaling are essential for liver specification in zebrafish, *Development*, *134*, 2041–2050.

[13] Ober, E.A., Verkade, H., Field, H.A., Stainier, D.Y. (2006). Mesodermal Wnt2b signalling positively regulates liver specification, *Nature*, *442*, 688–691.

[14] McLin, V.A., Rankin, S.A., & Zorn, A.M. (2007). Repression of Wnt/beta-catenin signaling in the anterior endoderm is essential for liver and pancreas development, *Development*, *134*, 2207–2217.

[15] Bort, R., Signore, M., Tremblay, K., Martinez Barbera, J.P., & Zaret, K.S. (2006). Hex homeobox gene controls the transition of the endoderm to a pseudostratified, cell emergent epithelium for liver bud development, *Developmental Biology*, *290*, 44–56.

[16] Lemaigre, F.P. (2009). Mechanisms of liver development: concepts for understanding liver disorders and design of novel therapies. *Gastroenterology*, *137*, 62-79.

[17] Tanimizu, N., & Miyajima, A. (2004). Notch signaling controls hepatoblast differentiation by altering the expression of liver enriched transcription factors, *Journal of Cell Science*, *117*, 3165–3174.

[18] Matsumoto, K., Miki, R., Nakayama, Tatsumi, N., & Yokouchi, Y. (2008). Wnt9a secreted from the walls of hepatic sinusoids is essential for morphogenesis, proliferation, and glycogen accumulation of chick hepatic epithelium, *Developmental Biology*, *319*, 234–247.

[19] Germain, L., Blouin, M.-J., & Marceau, N. (1988). Biliary epithelial and hepatocytic cell lineage relationships in embryonic rat liver as determined by the differential expression of cytokeratins, alpha-fetoprotein, albumin, and cell surface-exposed components, *Cancer Research*, *48*, 4909–4918.

[20] Suzuki, A., Iwana, A., Miyashita, H., Nakauchi, H., & Taniguchi, H. (2003) Role for growth factors and extracellular matrix in controlling differentiation of prospectively isolated hepatic stem cells, *Development*, *130*, 2513–2524.

[21] Terrace, J.D., Currie, I.S., Hay, D.C., Masson, N.M, Anderson, R.A, Forbes, S.J., Parks, R.W., & Ross, J.A. (2007). Progenitor cell characterization and location in the developing human liver, *Stem Cells and Development*, *16*, 771–778.

[22] Inada, M., Benten, B., Cheng, K., Joseph, B., Berishvili, E., Badve, S., Logdberg, L., Dabeva, M., & Gupta, S. (2008). Stage-specific regulation of adhesion molecule expression segregates epithelial stem/progenitor cells in fetal and adult human livers, *Hepatology International*, *2*, 50–62.

[23] Schmelzer, E., Zhang, L., Bruce, A., Wauthier, E., Ludlow, J., Yao, H.L., Moss, N., Melhem, A., McClelland, R., Turner, W., Kulik, M., Sherwood, S., Tallheden, T., Cheng, N., Furth, M.E., & Reid, L.M. (2007). Human hepatic stem cells from fetal and postnatal donors, *Journal of Experimental Medicine*, 204: 1973–1987.

[24] Rao, M.S., Khan, A.A., Parveen, N., Habeeb, M.A., Habibullah, C.M., & Pand G. (2008). Characterization of hepatic progenitors from human fetal liver during second trimester, World *Journal of Gastroenterology*, *14*, 5730-5737.

[25] Yovchev, M.I., Grozdanov, P.N., Zhou, H., Racherla, H., Guha, C., & Dabeva, M.D. (2008). Identification of adult hepatic progenitor cells capable of repopulating injured rat liver, *Hepatology*, *47*, 636-647.

[26] Sasaki, K., Kon, J., Mizuguchi, T., Chen, Q., Ooe, H., Oshima, H., Hirata, K., & Mitaka, T. (2008). Proliferation of hepatocyte progenitor cells isolated from adult human livers in serum-free medium, *Cell Transplantation*, *17*, 1221-1230.

[27] Yamashita, T., Budhu, A., Forgues, M., & Wang, X.M. (2007). Activation of hepatic stem cell marker EpCAM by Wnt-beta-catenin signaling in hepatocellular carcinoma, *Cancer Research*, *67*, 10831-10839.

[28] Zhang, L., Theise, N., Chua, M., & Reid, L.M. (2008). The stem cell niche of human livers: symmetry between development and regeneration, *Hepatology*, *48*, 1598-1607.

[29] Yamashita, T., Ji, J., Budhu, A., Forgues, M., Yang, W., Wang, H.Y., Jia, H., Ye, Q., Qin, L.X, Wauthier, E., Reid, L.M., Minato, H., Honda, M., Kaneko, S., Tang, Z.Y., & Wang, X.W. (2009). EpCAM-positive hepatocellular carcinoma cells are tumor-initiating cells with stem/progenitor cell features, *Gastroenterology*, *136*, 1012-1024.

[30] Terrace, J.D., Hay, D.C., Samuel, K., Payne, C., Anderson, R.A., Currie, I.S, Forbes, R.W., Parks, S.J., & Ross, J.A. (2009). Side population cells in developing human liver are primarily haematopoietic progenitor cells, *Experimental Cell Research*, *315*, 2141–2153.

[31] Goodell, M.A., McKinney-Freeman, S., & Camargo, F.D. (2005) Isolation and characterization of side population cells, *Methods in Molecular Biology*, *290*, 343–352.

[32] Dan, Y.Y., Riehle, K.J., Lazaro, C., Teoh, N., Haque, J., Campbell, J.S., & Fausto, N. (2006). Isolation of multipotent progenitor cells from human fetal liver capable of differentiating into liver and mesenchymal lineages, *Proceedings of the National Academy of Sciences of USA*, *103*, 9912-9917.

[33] Hu, C., Li, H., Li, J., Zhu, Z., Yin, S., Hao, X., Yao, M., Zheng, S., Gu, J. (2008). Analysis of ABCG2 expression and side population identifies intrinsic drug efflux in the HCC cell line MHCC-97L and its modulation by Akt signaling, *Carcinogenesis*, *29*, 2289-2297.

[34] Kamohara, Y., Haraguchi, N., Mimori, K., Tanaka, F., Inoue, H., Mori, M., & Kanematsu, T. (2008). The search for cancer stem cells in hepatocellular carcinoma, *Surgery*, *144*, 119-124.

[35] Shi, G.M., Xu, Y., Fan, J., Zhou, J., Yang, X.R., Qiu, S.J., Liao, Y., Wu, W.Z., Ji, Y., Ke, A.W., Ding, Z.B., He, Y.Z., Wu, B., Yang, G.H., Qin, W.Z., Zhang, W., Zhu, J., Min, Z.H., & Wu, Z.Q. (2008). Identification of side population cells in human hepatocellular carcinoma cell lines with stepwise metastatic potentials, *Journal of Cancer Research and Clinical Oncology*, *134*, 1155-1163.

[36] Fausto, N., & Campbell, J.S. (2003). The role of hepatocytes and oval cells in liver regeneration and repopulation, *Mechanisms of Development*, *120*, 117-130.

[37] Evarts, R.P., Nagy, P., Marsden, E., & Thorgeirsson, S.S. (1987). A precursor-product relationship exists between oval cells and hepatocytes in rat liver, *Carcinogenesis*, *8*, 1737-1740.

[38] Fujio, K., Evarts, R.P., Hu, Z., Marsden, R.P., & Thorgeirsson, S.S. (1994). Expression of stem cell factor and its receptor, c-kit, during liver regeneration from putative stem cells in adult rat. *Laboratory Investigation*, *70*, 511-516.

[39] Jensen, C.H., Jauho, E.I., Santoni-Rugiu, E., Holmskov, U., Teisner, B., Tygstrup, N., & Bisgaard, H.C. (2004). Transit-amplifying ductular (oval) cells and their hepatocytic progeny are characterized by a novel and distinctive expression of delta-like protein/preadipocyte factor 1/fetal antigen 1, *American Journal of Pathology*, *164*, 1347-1359.

[40] Petersen, B.E., Goff, J.P., Greenberger, J.S., & Michalopoulos, G.K. (1998). Hepatic oval cells express the hematopoietic stem cell marker Thy-1 in the rat, *Hepatology*, *27*, 433-445. Petersen, B.E., Bowen, W.C., Patrene, K.D., Mars, W.M., Sullivan, A.K., Murase, N., Boggs, S.S., Greenberger, J.S., & Goff, J.P. (1999). Bone marrow as a potential source of hepatic oval cells, *Science*, *284*, 1168-1170.

[41] Wang, X., Foster, M., Al-Dhalimy, M., Lagasse, E., Finegold, M., & Grompe, M. (2003). The origin and liver repopulating capacity of murine oval cells, *Proceedings of the National Academy of Sciences of USA*, *100*, 11881-11888.

[42] Thorgeirsson, S.S., & Grisham, J.W. (2006). Hematopoietic cells as hepatocyte stem cells: a critical review of the evidence, *Hepatology*, *43*, 2-8.

[43] Oh, S.H., Witek, R.P., Bae, S.H., Zheng, D., Jung, Y., Piscaglia, A.C., & Petersen, B.E. (2007). Bone marrow-derived hepatic oval cells differentiate into hepatocytes in 2-acetylaminofluorene/ partial hepatectomy-induced liver regeneration, *Gastroenterology*, *132*, 1077-1087.

[44] Oertel, M., & Shafritz, D.A. (2008). Stem cells, cell transplantation and liver repopulation, *Biochimica et Biophysica Acta*, *1782*, 61-74.

[45] Fausto, N. (2000). Liver regeneration, *Journal of Hepatology*, *32*, 19-31.

[46] Wang, P., Cong, M., Liu, T.H., Yang, A.T., Cong, R., Wu, P., Tang, S.Z., Xu, Y., Wang, H., Wang, B.E., Jia, J.D., & You, H. (2010). Primary isolated hepatic oval cells maintain progenitor cell phenotypes after two-year prolonged cultivation, *Journal of Hepatology*, *53*, 863-871.

[47] Grisham, J.W., Coleman, W.B., & Smith, G.J. (1993). Isolation, culture, and transplantation of rat hepatocytic precursor (stem-like) cells, *Proceedings of the Society for Experimental Biology and Medicine*, *204*, 270-279.

[48] Coleman, W.B., McCullough, K.D., Esch, G.L., Faris, R.A., Hixson, D.C., Smith, G.J, & Grisham, G.J. (1997). Evaluation of the differentiation potential of WB-F344 rat liver epithelial stem-like cells in vivo. Differentiation to hepatocytes after transplantation into dipeptidylpeptidase-IV-deficient rat liver, *American Journal of Pathology*, *151*, 353-359.

[49] Dabeva, M.D., Hwang, S.G., Vasa, S.R., Hurston, E., Novikoff, P.M., Hixson, D.C., Gupta, S., & Shafritz, D.A. (1997). Differentiation of pancreatic epithelial progenitor cells into hepatocytes following transplantation into rat liver, *Proceedings of the National Academy of Sciences of USA*, *94*, 7356-7361.

[50] Song, S., Witek, R.P., Lu, Y., Choi, Y.K., Zheng, D., Jorgensen, M., Li, C., Flotte, T.R., Petersen, B.E. (2004). Ex vivo transduced liver progenitor cells as a platform for gene therapy in mice. *Hepatology*, *40*, 918-924.

[51] Kisseleva, T., Gigante, E., & Brenner, D.A. (2010). Recent advances in liver stem cell therapy, *Current Opinion in Gastroenterology*, *26*, 395-402.

[52] Mitaka, T., Kojima, T., Mizuguchi, T., & Mochizuki, Y. (1995). Growth and maturation of small hepatocytes isolated from adult rat liver, *Biochemical and Biophysical Research Communications*, *214*, 310-317.

[53] Ikeda, S., Mitaka, T., Harada, K., Sugimoto, S., Hirata, K., & Mochizuki, Y. (2002). Proliferation of rat small hepatocytes after long-term cryopreservation, *Journal of Hepatology*, *37*, 7-14.

[54] Sidler Pfandler, M.A., Hochli, M., Inderbitzin, D., Meier, P.J., & Stieger, B. (2004). Small hepatocytes in culture develop polarized transporter expression and differentiation. *Journal of Cell Science*, *117*, 4077-4087.

[55] Mitaka, T., Mizuguchi, T., Sato, F., Mochizuki, C., & Mochizuki, Y. (1998). Growth and maturation of small hepatocytes, *Journal of Gastroenterology and Hepatology*, *13*, S70-S77.

[56] Mitaka, T., & Ooe, H. (2010). Characterization of hepatic-organoid cultures, *Drug Metabolism Reviews*, *42*, 472-481.

[57] Michalopoulos, G.K., Bowen, W.C., Mulè, K., & Stolz, D.B. (2001). Histological organization in hepatocyte organoid cultures, American Journal Pathology, *159*, 1877-1887.

[58] Wu, M.H., Huang, S.B., & Lee, G.B. (2010). Microfluidic cell culture systems for drug research, *Lab on a Chip*, *10*, 939-956.

[59] Herrera, M.B., Bruno, S., Buttiglieri, S., Tetta, C., Gatti, S., Deregibus, M.C., Bussolati, B., & Camussi, G. (2006). Isolation and characterization of a stem cell population from adult human liver, *Stem Cells*, *24*, 2840-2850.

[60] Khuu, D.N., Najimi, M., & Sokal, E.M. (2007). Epithelial cells with hepatobiliary phenotype: is it another stem cell candidate for healthy adult human liver? *World Journal of Gastroenterology*, *13*, 1554-1560.

[61] Duret, C., Gerbal-Chaloin, S., Ramos, J., Fabre, J.M., Jacquet, E., Navarro, F., Blanc, P., Sa-Cunha, A., Maurel, P., & Daujat-Chavanieu, M. (2007). Isolation, characterization, and differentiation to hepatocyte-like cells of nonparenchymal epithelial cells from adult human liver, *Stem Cells*, *25*, 1779-1790.

[62] Castorina, S., Torrisi, T.A., Privitera, G., & Panebianco, M. (2008). Isolation of epithelial cells with hepatobiliary phenotype, *Italian Journal of Anatomy and Embryology*, *113*, 199-207.

[63] Najimi, M., Khuu, D.N., Lysy, P.A., Jazouli, N., Abarca, J., Sempoux, C., & Sokal, E.M. (2007). Adult derived human liver mesenchymal-like cells as a potential progenitors reservoir of hepatocytes? *Cell Transplantation*, *16*, 717-728.

[64] Tarnowski , M., Koryciak-Komarska, H., Czekaj, P., Sebesta, R., Czekaj, T.M., Urbanek, K., Likus, W., Malinowska-Kolodziej, I., Plewka, D., Nowaczyk-Dura, G., & Wiaderkiewicz, R. (2007). The comparison of multipotential for differentiation of progenitor mesenchymal-like stem cells obtained from livers of young and old rats, *Folia Histochemica et Cytobiologica*, *45*, 245-254.

[65] Meng, Z., Fu, X., Chen, X., Zeng, S., Tian, Y., Jove, R., Xu, R., Huang, W. (2010). miR-194 is a marker of hepatic epithelial cells and suppresses metastasis of liver cancer cells in mice, *Hepatology*, *52*, 2148-2157.

[66] Karbanová, J., & Mokrý, J. (2002). Histological and histochemical analysis of embryoid bodies, *Acta Histochemica*, *104*, 361-365.

[67] Zhou, Q.J., Shao, J.Z., Xiang, L.X., Hu, R.Z., Lu, Y.L., Yao, H., Dai, L.C. (2005). Generation of embryoid bodies from mouse embryonic stem cells cultured on STO feeder cells, *Cell Biology International*, *29*, 817-825.

[68] Karp, J.M., Ferreira, L.S., Khademhosseini, A., Kwon, A.H., Yeh, J., & Langer, R.S. (2006). Cultivation of human embryonic stem cells without the embryoid body step enhances osteogenesis in vitro, *Stem Cells*, *24*, 835–843.

[69] Choi, D., Lee, H.J., Jee, S., Jin, S., Koo, S.K., Paik, S.S., Jung, S.C., Hwang, S.Y., Lee, K.S., Oh, B. (2005). In vitro differentiation of mouse embryonic stem cells: enrichment of endodermal cells in the embryoid body, *Stem Cells*, *23*, 817-827.

[70] Chinzei, R., Tanaka, Y., Shimizu-Saito, K., Hara, Y., Kakinuma, S., Watanabe, M., Teramoto, K., Arii, S., Takase, K., Sato, C., Terada, N., & Teraoka, H. (2002). Embryoid-body cells derived from a mouse embryonic stem cell line show differentiation into functional hepatocytes, *Hepatology*, *36*, 22–29.

[71] Asahina, K., Fujimori, H., Shimizu-Saito, K., Kumashiro, Y., Okamura, K., Tanaka, Y., Teramoto, K., Arii, S., & Teraoka, H. (2004). Expression of the liver-specific gene Cyp7a1 reveals hepatic differentiation in embryoid bodies derived from mouse embryonic stem cells. *Genes Cells*, *9*, 1297–1308.

[72] Hu, A.B., Cai, J.Y., Zheng, Q.C., He, X.Q., Shan, Y., Pan, Y.L., Zeng, G.C., Hong, A., Dai, Y., Li, L.S. (2004). High-ratio differentiation of embryonic stem cells into hepatocytes in vitro. *Liver International*, *24*, 237-245.

[73] Gouon-Evans, V., Boussemart, L., Gadue, P., Nierhoff, D., Koehler, C.I., Kubo, A., Shafritz, D.A., & Keller, G. (2006). BMP-4 is required for hepatic specification of mouse embryonic stem cell-derived definitive endoderm, *Nature Biotechnology*, *24*, 1402-1411.

[74] Hay, D.C., Fletcher, J., Payne, C., Terrace, J.D., Gallagher, R.C., Snoeys, J., Black, J.R., Wojtacha, D., Samuel, K., Hannoun, Z., Pryde, A., Filippi, C., Currie, I.S., Forbes, S.J., Ross, J.A., Newsome, P.N., & Iredale, J.P. (2008). Highly efficient differentiation of hESCs to functional hepatic endoderm requires Activin A and Wnt3a signaling. *Proceedings of the National Academy Sciences of USA*, *105*, 12301–12306.

[75] Soto-Gutierrez, A., Navarro-Alvarez, N., Caballero-Corbalan, J., Tanaka, N., & Kobayashi, N. (2008). Endoderm induction for hepatic and pancreatic differentiation of ES cells, *Acta Medica Okayama*, *62*, 63-68.

[76] Sekhon, S.S., Tan, X., Micsenyi, A., Bowen, W.C., & Monga, S.P. (2004). Fibroblast growth factor enriches the embryonic liver cultures for hepatic progenitors. *American Journal of Pathology*, *164:* 2229-2240.

[77] Roelandt, P., Pauwelyn, K.A., Sancho-Bru, P., Subramanian, K., Ordovas, L., Vanuytsel, K., Geraerts, M., Firpo, M., De Vos, R., Fevery, J., Nevens, F., Hu, W.S., & Verfaillie, C.M. (2010). Human embryonic and rat adult stem cells with primitive endoderm-like phenotype can be fated to definitive endoderm, and finally hepatocyte-like cells, *PLoS One*, *5*: e12101.

[78] Touboul, T., Hannan, N.R., Corbineau, S., Subramanian, K., Ordovas, L., Vanuytsel, K., Geraerts, M., Firpo, M., De Vos, R., Fevery, J., Nevens, F., Hu, W.S., & Verfaillie, C.M. (2010). Generation of functional hepatocytes from human embryonic stem cells under chemically defined conditions that recapitulate liver development, *Hepatology*, *51*, 1754-1765.

[79] Lavon, N. (2010). Generation of hepatocytes from human embryonic stem cells, *Methods in Molecular Biology*, *640*, 237-246.

[80] Ma, X., Duan, Y., Jung, C.J., Wu, J., VandeVoort, C.A., & Zern, M.A. (2008). The differentiation of hepatocyte-like cells from monkey embryonic stem cells, *Cloning Stem Cells*, *10*, 485-493.

[81] Weigel, P.H. (1980). Characterization of the asialoglycoprotein receptor on isolated rat hepatocytes. *Journal of Biological Chemistry*, *255*, 6111-6120.

[82] Basma, H., Soto-Gutiérrez, A., Yannam, G.R., Liu, L., Ito, R., Yamamoto, T., Ellis, E., Carson, S.D., Sato, S., Chen, Y., Muirhead, D., Navarro-Alvarez, N., Wong, R.J., Roy-Chowdhury, J., Platt, J.L., Mercer, D.F., Miller, J.D., Strom, S.C., Kobayashi, N., & Fox, I.J. (2009). Differentiation and transplantation of human embryonic stem cell-derived hepatocytes, *Gastroenterology*, *136*, 990-999.

[83] Baharvand, H., Hashemi, S.M., Kazemi Ashtiani, S., & Farrokhi, A. (2006). Differentiation of human embryonic stem cells into hepatocytes in 2D and 3D culture systems in vitro, *International Journal of Developmental Biology*, *50*: 645-652.

[84] Liu, T., Zhang, S., Chen, X., Li, G., & Wang, Y. (2010). Hepatic differentiation of mouse embryonic stem cells in three-dimensional polymer scaffolds, *Tissue Engineering Part A*, *16*, 1115-1122.

[85] Anderson, D.G., Burdick, J.A., & Langer, R. (2004). Smart biomaterials, *Science*, *305*, 1923–1924.

[86] Wang, S., Nagrath, D., Chen, P.C., Berthiaume, F., & Yarmush, M.L. (2008). Three-dimensional primary hepatocyte culture in synthetic self-assembling peptide hydrogel, *Tissue Engineering Part A*, *14*, 227-236.

[87] Semino, C.E., Merok, J.R., Crane, G.G., Panagiotakos, G., Zhang, S. (2003). Functional differentiation of hepatocyte-like spheroid structures from putative liver progenitor cells in three-dimensional peptide scaffolds, *Differentiation*, *71*, 262-270.

[88] Lee, K.D., Kuo, T.K., Whang-Peng, J., Chung, Y.F., Lin, C.T., Chou, S.H., Chen, J.R., Chen, Y.P., & Lee, O.K. (2004). In vitro hepatic differentiation of human mesenchymal stem cells, *Hepatology*, *40*, 1275-1284.

[89] Zemel, R., Bachmetov, L., Ad-El, D., Abraham, A., & Tur-Kaspa, R. (2009). Expression of liver-specific markers in naïve adipose-derived mesenchymal stem cells, *Liver International*, *29*, 1326-1337.

[90] Seo, M.J., Suh, S.Y., Bae, Y.C., & Jung, J.S. (2005). Differentiation of human adipose stromal cells into hepatic lineage in vitro and in vivo, *Biochemical and Biophysical Research Communications*, *328*, 258-264.

[91] Aurich, H., Sgodda, M., Kaltwasser, P., Vetter, M., Weise, A., Liehr, T., Brulport, M., Hengstler, J.G., Dollinger, M.M., Fleig, W.E., & Christ, B. (2009). Hepatocyte differentiation of mesenchymal stem cells from human adipose tissue in vitro promotes hepatic integration in vivo, Gut, 58, 570-581.

[92] Cantz, T., Sharma, A.D., Jochheim-Richter, A., Arseniev, L., Klein, C., Manns, M.P., & Ott, M. (2004). Reevaluation of bone marrow-derived cells as a source for hepatocyte regeneration, *Cell Transplantation*, *13*, 659-666.

[93] Aurich, I., Mueller, L.P., Aurich, H., Luetzkendorf, J., Tisljar, K., Dollinger, M.M., Schormann, W., Walldorf, J., Hengstler, J.G., Fleig, W.E., & Christ, B. (2007). Functional integration of hepatocytes derived from human mesenchymal stem cells into mouse livers, *Gut*, *56*, 405-415.

[94] Kogler, G., Sensken, S., Airey, J.A., Trapp, T., Muschen, M., Feldhahn, N., Liedtke, S., Sorg, R.V., Fischer, J., Rosenbaum, C., Greschat, S., Knipper, A., Bender, J., Degistirici, O., Gao, J., Caplan, A.I., Colletti, E.J., Almeida-Porada, G., Muller, H.W., Zanjani, E., & Wernet P. (2004). A new human somatic stem cell from placental cord blood with intrinsic pluripotent differentiation potential, *Journal of Experimental Medicine*, *200*, 123-135.

[95] Sato, Y., Araki, H., Kato, J., Nakamura, K., Kawano, Y., Kobune, M., Sato, T., Miyanishi, K., Takayama, T., Takahashi, M., Takimoto, R., Iyama, S., Matsunaga, T., Ohtani, S., Matsuura, A., Hamada, H., & Niitsu, Y.(2005). Human mesenchymal stem cells xenografted directly to rat liver are differentiated into human hepatocytes without fusion, Blood, *106*, 756-763.

[96] Mohamadnejad, M., Alimoghaddam, K., Mohyeddin-Bonab, M., Bagheri, M., Bashtar, M., Ghanaati, H., Baharvand, H., Ghavamzadeh, A., & Malekzadeh, R. (2007). Phase 1 trial of autologous bone marrow mesenchymal stem cell transplantation in patients with decompensated liver cirrhosis, *Archives of Iranian Medicine*, *10*, 459-466.

[97] Kharaziha, P., Hellström, P.M., Noorinayer, B., Farzaneh, F., Aghajani, K., Jafari, F., Telkabadi, M., Atashi, A., Honardoost, M., Zali, M.R., & Soleimani, M. (2009). Improvement of liver function in liver cirrhosis patients after autologous mesenchymal stem cell injection: a phase I-II clinical trial, *European Journal of Gastroenterology and Hepatology*, *21*, 1199-1205.

[98] Prasad, V.K., Lucas, K.G., Kleiner, G.I., Talano, J.A., Jacobsohn, D., Broadwater, G., Monroy, R., & Kurtzberg, J. (2011). Efficacy and safety of ex-vivo cultured adult human mesenchymal stem cells (Prochymal™) in pediatric patients with aevere refractory acute graft-versus-host disease in a compassionate use study, *Biology of Blood and Marrow Transplantation*, *17*, 534-541.

[99] Wang, X., Ge, S., McNamara, G., Hao, Q.L., Crooks, G.M., & Nolta, J.A. (2003). Albumin-expressing hepatocyte-like cells develop in the livers of immune-deficient mice that received transplants of highly purified human hematopoietic stem cells, *Blood*, *101*, 4201-4208.

[100] Wang, X., Montini, E., Al-Dhalimy, M., Lagasse, E., Finegold, M., Grompe, M. (2002). Kinetics of liver repopulation after bone marrow transplantation, *American Journal of Pathology*, *161*, 565-574.

[101] Lagasse, E., Connors, H., Al-Dhalimy, M., Reitsma, M., Dohse, M., Osborne, L., Wang, X., Finegold, M., Weissman, I.L., & Grompe, M. (2000). Purified hematopoietic stem cells can differentiate into hepatocytes in vivo, *Nature Medicine*, *6*, 1229-1234.

[102] Jang, Y.Y., Collector, M.I., Baylin, S.B., Diehl, A.M., & Sharkis, S.J. (2004). Hematopoietic stem cells convert into liver cells within days without fusion, *Nature Cell Biology*, *6*, 532-539.

[103] Terai, S., Ishikawa, T., Omori, K., Aoyama, K., Marumoto, Y., Urata, Y., Yokoyama, Y., Uchida, K., Yamasaki, T., Fujii, Y., Okita, K., & Sakaida, I. (2006). Improved liver function in patients with liver cirrhosis after autologous bone marrow cell infusion therapy, *Stem Cells*, *24*, 2292-2298.

[104] Lyra, A.C., Soares, M.B., da Silva, L.F., Fortes, M.F., Silva, A.G., Mota, A.C., Oliveira, S.A., Braga, E.L., de Carvalho, W.A., Genser, B., dos Santos, R.R., & Lyra, L.G. (2007). Feasibility and safety of autologous bone marrow mononuclear cell transplantation in patients with advanced chronic liver disease, *World Journal of Gastroenterology*, *13*, 1067-1073.

[105] Pai, M., Zacharoulis, D., Milicevic, M.N., Helmy, S., Jiao, L.R., Levicar, N., Tait, P., Scott, M., Marley, D.B., Jestice, K., Glibetic, M., Bansi, D., Khan, S.A., Kyriakou, D., Rountas, C., Thillainayagam,A., Nicholls, J.P., Jensen, S., Apperley, J.F., Gordon, M.Y., & Habib, N.A. (2008). Autologous infusion of expanded mobilized adult bone marrow-derived CD34+ cells into patients with alcoholic liver cirrhosis, *American Journal of Gastroenterology*, *103,* 1952-1958.

[106] Levicar, N., Pai, M., Habib, N.A., Tait, P., Jiao, L.R., Marley, S.B., Davis, J., Dazzi, F., Smadja, C., Jensen, S.L., Apperley, J.P.F., & Gordon, M.Y. (2008). Long-term clinical results of autologous infusion of mobilized adult bone marrow derived CD34+ cells in patients with chronic liver disease, *Cell Proliferation*, *41*, 115-125.

[107] Lorenzini, S., Isidori, A., Catani, L., Gramenzi, A., Talarico, S., Bonifazi, F., Giudice, V., Conte, R., Baccarani, M., Bernardi, M., Forbes, S.J., Lemoli, R.M., & Andreone, P. (2008). Stem cell mobilization and collection in patients with liver cirrhosis, *Alimentary Pharmacology and Therapy*, *27*, 932-939.

[108] di Bonzo, L.V., Ferrero, I., Cravanzola, C., Mareschi, K., Rustichell, D., Novo, E., Sanavio, F., Cannito, S., Zamara, E., Bertero, M., Davit, A., Francica, S., Novelli, F., Colombatto, S., Fagioli, F., & Parola, M. (2008). Human mesenchymal stem cells as a two-edged sword in hepatic regenerative medicine: engraftment and hepatocyte differentiation versus profibrogenic potential, *Gut*, *57*, 223-231.

[109] Dalakas, E., Newsome, P.N., Boyle, S., Brown, R., Pryde, A., McCall, S., Hayes, P.C., Bickmore, W.A., Harrison, D.J., & Plevris, J.N. (2010). Bone marrow stem cells contribute to alcohol liver fibrosis in humans, *Stem Cells and Developement, 19*, 1417-1425.

[110] Aoi, T., Ya, K., Nakagawa, M., Ichisaka, T., Okita, K., Takahashi, K., Chiba, T., & Yamanaka, S. (2008). Generation of pluripotent stem cells from adult mouse liver and stomach cells, *Science, 321*, 699-702.

[111] Liu, H., Ye, Z., Kim, Y., Sharkis, S., & Jang, Y.Y. (2010). Generation of endoderm-derived human induced pluripotent stem cells from primary hepatocytes, *Hepatology, 51*, 1810-1819.

[112] Song, Z., Cai, J., Liu, Y., Zhao, D., Yong, J., Duo, S., Song, X., Guo, Y., Zhao, Y., Qin, H., Yin, X., Wu, C., Che, J., Lu, S., Ding, M., & Deng, H. (2009). Efficient generation of hepatocyte-like cells from human induced pluripotent stem cells, *Cell Research, 19*, 1233-1242.

[113] Si-Tayeb, K., Noto, F.K., Nagaoka, M., Li, J., Battle, M.A., Duris, C., North, P.E., Dalton, S., & Duncan, S.A. (2010). Highly efficient generation of human hepatocyte-like cells from induced pluripotent stem cells, *Hepatology, 51*, 297-305.

[114] Gai, H., Nguyen, D.M., Moon, Y.J., Aguila, J.R., Fink, L.M., Ward, D.C., & Ma, Y. (2010). Generation of murine hepatic lineage cells from induced pluripotent stem cells, *Differentiation, 79*, 171-181.

[115] Iwamuro, M., T. Komaki, Y. Kubota, M. Seita, H. Kawamoto, T. Yuasa, J.M. Shahid, R.A. Hassan, W.A. Hassan, S. Nakaji, Y. Nishikawa, E. Kondo, K. Yamamoto, I.J., & Fox, N. (2010). Hepatic differentiation of mouse iPS cells in vitro, *Cell Transplantation, 19*, 841-847.

[116] Polo, J.M., Liu, S., Figueroa, M.E., Kulalert, M., Eminli, S., Tan, K.Y., Apostolou, E., Stadtfeld, M., Li, Y., Shioda, T., Natesan, S., A.J. Wagers, Melnick, A., Evans, T., & Hochedlinger, K. (2010). Cell type of origin influences the molecular and functional properties of mouse induced pluripotent stem cells, *Nature Biotechnology, 28*, 848-855.

[117] Sancho-Bru, P., Roelandt, P., Narain, N., Pauwelyn, K., Notelaers, K., Shimizu, T., Ott, M., & Verfaillie, C. (2011). Directed differentiation of murine-induced pluripotent stem cells to functional hepatocyte-like cells. *Journal of Hepatology, 54*, 98-107.

[118] Denker, H.W. (2009). Induced pluripotent cells: how to deal with the developmental potential, *Reproductive Biomedicine Online, 19*, 34-37.

[119] Hong, H., Takahashi, K., Ichisaka, T., Aoi, T., Kanagawa, O., Nakagawa, M., Okita, K., & Yamanaka, S. (2009). Suppression of induced pluripotent stem cell generation by the p53-p21 pathway, *Nature, 460*, 1132–1135.

[120] Utikal, J., Polo, J.M., Stadtfeld, M., Maherali, N., Kulalert, W., Walsh, R.M., Khalil, A., Rheinwald, J.G., & Hochedlinger, K. (2009). Immortalization eliminates a roadblock during cellular reprogramming into iPS cells, *Nature, 460*, 1145–1148.

[121] Okita, K., Nakagawa, M., Hyenjong, H., Ichisaka, T., & Yamanaka, S. (2008). Generation of mouse induced pluripotent stem cells without viral vectors, *Science, 322:* 949–953.

[122] Kaji, K., Norrby, K., Paca, A., Mileikovsky, M., Mohseni, P., & Woltjen, K. (2009). Virus-free induction of pluripotency and subsequent excision of reprogramming factors, *Nature, 458*, 771-775.

[123] Woltjen, K., Michael, I.P., Mohseni, P., Desai, R., Mileikovsky, M., Hämäläinen, R., Cowling, R., Wang, W., Liu, P., Gertsenstein, M., Kaji, K., Sung, H.K., & Nagy, A. (2009). piggyBac transposition reprograms fibroblasts to induced pluripotent stem cells, *Nature*, *458*, 766-770.

[124] Kim, J.B., Sebastiano, V., Wu, G., Araúzo-Bravo, M.J., Sasse, P., Gentile, L., Ko, K., Ruau, D., Ehrich, M., van den Boom, D., Meyer, J., Hübner, K., Bernemann, C., Ortmeier, C., Zenke, M., Fleischmann, B.K., Zaehres, H., & Schöler, H.R. (2009). Oct4-induced pluripotency in adult neural stem cells, *Cell*, *136:* 411-419.

[125] Huang, P., He, Z., Ji, S., Sun, H., Xiang, D., Liu, C., Hu, Y., Wang, X., & Hui, L. (2011) Induction of functional hepatocyte-like cells from mouse fibroblasts by defined factors, *Nature*, *475*, 386-389.

[126] Ellor, S., Shupe, T., & Petersen, B. (2008). Stem cell therapy for inherited metabolic disorders of the liver, *Experimental Hematology*, *36*, 716-725.

[127] Grossman, M., Raper, S.E., Kozarsky, K., Stein, E.A., Engelhardt, J.F., Muller, D., Lupien, P.J., & Wilson, J.M. (1994). Successful ex vivo gene therapy directed to liver in a patient with familial hypercholesterolaemia, *Nature Genetics*, *6*, 335-341.

[128] Fisher, R.A., & Strom, S.C. (2006). Human hepatocyte transplantation: worldwide results, *Transplantation*, *82*, 441-449.

[129] Allen, K.J., & Soriano, H.E. (2001). Liver cell transplantation: The road to clinical application, *Journal of Laboratory and Clinical Medicine*, *138*, 298-312.

[130] Horslen, S.P., & Fox, I.J. (2004). Hepatocyte transplantation, *Transplantation*, *77*, 1481-1486.

[131] Mitry, R.R., Dhawan, A., Hughes, R.D., Bansal, S., Lehec, S., Terry, C., Heaton, N.D., Karani, J.B., Mieli-Vergani, G., & Rela, M. (2004). One liver, three recipients: Segment IV from split-liver procedures as a source of hepatocytes for cell transplantation, Transplantation, *77*, 1614-1616.

[132] Strom, S.C., Chowdhury, J.R., & Fox, I.J. (1999). Hepatocyte transplantation for the treatment of human disease, *Seminars in Liver Disease*, *19*, 39-48.

[133] Bilir, B., Durham, J., Krystal, J., Karrer, F., Kumpe, D., & Ostrowska, A. (1996). Transjugular intra-portal transplantation of cryopreserved human hepatocytes in a patient with acute liver failure, *Hepatology*, *24*, 308A.

[134] Soriano, H., Wood, R., Kang, D., Ozaki, C., Finegold, M., & Bishoff, F. (1997). Hepatocellular transplantation in children with fulminant liver failure, *Hepatology*, *26*, 239A.

[135] Fisher, R.A., Bu, D., Thompson, M., Tisnado, J., Prasad, U., & Sterling, R. (2000). Defining hepatocellular chimerism in a liver failure patient bridged with hepatocyte infusion, *Transplantation*, *69*, 303-307.

[136] Bilir, B.M., Guinette, D., Karrer, F., Kumpe, D.A., Krysl, J., Stephens, J., McGavran, L., Ostrowska, A., & Durham, J. (2000). Hepatocyte transplantation in acute liver failure. *Liver Transplantation*, *6*, 32-40.

[137] Smets, F., Najimi, M., & Sokal, E.M. (2008). Cell transplantation in the treatment of liver diseases, *Pediatric Transplantation*, *12*, 6-13.

[138] Trono, D. (2000). Lentiviral vectors: turning a deadly foe into a therapeutic agent, *Gene Therapy*, *7*, 20-23.

[139] Nguyen, T.H., Oberholzer, J., Birraux, J., Majno, P., Morel, P., & Trono, D. (2002). Highly efficient lentiviral vector-mediated transduction of nondividing, fully reimplantable primary hepatocytes, *Molecular Therapy*, *6*: 199-209.

[140] Azuma, H., Paulk, N., Ranade, A., Dorrell, C., Al-Dhalimy, A., Ellis, E., Strom, S., Kay, M.A., Finegold, M., & Grompe, M. (2007). Robust expansion of human hepatocytes in Fah-/-/Rag2-/-/Il2rg-/- mice. *Nature Biotechnology*, *25*, 903-910.

[141] Malhi, H., Joseph, B., Schilsky, M.L., & Gupta, S. (2008). Development of cell therapy strategies to overcome copper toxicity in the LEC rat model of Wilson disease, *Regenerative Medicine*, *3*, 165-173.

[142] Yokoyama, T., Ohashi, K., Kuge, H., Kanehiro, H., Iwata, H., Yamato, M., & Nakajima, Y. (2006). In vivo engineering of metabolically active hepatic tissues in a neovascularized subcutaneous cavity, *American Journal of Transplantation*, *6*: 50-59.

[143] Ohashi, K., Yokoyama, T., Yamato, M., Kuge, H., Kanehiro, H., Tsutsumi, M., Amanuma, T., Iwata, H., Yang, J., Okano, T., & Nakajima, Y. (2007). Engineering functional two- and three-dimensional liver systems in vivo using hepatic tissue sheets, *Nature Medicine*, *13*: 880-885.

[144] van de Kerkhove, M.P., Hoekstra, R., Chamuleau, R.A.F.M., & van Gulik, T.M. (2004). Clinical application of bioartificial liver support systems, *Annals of Surgery*, *240*, 216–230.

[145] Naruse, K. (2005). Artificial liver support: future aspects, *Journal of Artificial Organs*, *8*, 71–76.

[146] Fox, I.J., Langnas, A.N., Fristoe, L.W., Shaefer, M.S., Vogel, J.E., Antonson, D.L., Donovan, J.P., Heffron, T.G., Markin, R.S., & Sorrell, M.F. (1993). Successful application of extracorporeal liver perfusion: a technology whose time has come, *American Journal of Gastroenterology*, *88*, 1876–1881.

[147] Xiong, A., Austin, T.W., & Lagasse, E. (2008). Isolation of human fetal liver progenitors and their enhanced proliferation by three-dimensional coculture with endothelial cells, *Tissue Engineering Part A*, *14*, 995–1006.

[148] van de Kerkhove, M.P., Poyck, P.P.C., Deurholt, T., Hoekstra, R., Chamuleau, R.A.F.M., & van Gulik, T.M. (2005). Liver support therapy: an overview of the AMC-bioartificial liver research, *Digestive Surgery*, *22*, 254–264.

[149] Fruhauf, J.-H., Mertsching, H., Giri, S., Fruhauf, N.R., & Bader, R. (2009). Porcine endogenous retrovirus released via bioartificial liver infects primary human cells, *Liver International*, *29*, 1553–1561.

[150] Chamuleau, R.A.F.M., Poyck, P.P.C., & van de Kerkhove, M.-P. (2006). Bioartificial liver: its pros and cons, *Therapeutic Apheresis and Dialysis*, *10*, 168–174.

[151] Sgroi, A., Serre-Beinier, V., Morel, P., & Buhler, L. (2009). What clinical alternatives to whole liver transplantation? Current status of artificial devices and hepatocyte transplantation, *Transplantation*, *87*: 457–466.

[152] Sauer, I.M., & Gerlach, J.C. (2002). Modular extracorporeal liver support, *Artificial Organs*, *26*, 703-706.

[153] Miyazawa, M., Torii, T., Toshimitsu, Y., Okada, K., & Koyama, I. (2007). Hepatocyte dynamics in a three-dimensional rotating bioreactor, *Journal of Gastroenterology and Hepatology*, *22*: 1959–1964.

[154] Sussman, N.L., Gislason, G.T., Conlin, C.A., & Kelly, J.H. (1994). The Hepatix extracorporeal liver assist device: initial clinical experience, *Artificial Organs*, *18*, 390-396.

[155] Sussman, N.L., & Kelly, J.H. (1997). Extracorporeal liver support: cell-based therapy for the failing liver, *American Journal of Kidney Diseases*, *30*, S66-S71.

[156] Kosuge, M., Takizawa, H., Maehashi, H., Matsuura, T., & Matsufuji, S. (2007). A comprehensive gene expression analysis of human hepatocellular carcinoma cell lines as components of a bioartificial liver using a radial flow bioreactor, *Liver International*, *27*, 101–108.

[157] Poyck, P.P., van Wijk, A.C., van der Hoeven, T.V., de Waart, D.R., Chamuleau, R.A., van Gulik, T.M., Oude Elferink, R.P., & Hoekstra, R. (2008). Evaluation of a new immortalized human fetal liver cell line (cBAL111) for application in bioartificial liver, *Journal of Hepatology*, *48*, 266–275.

[158] Patzer, J.F., 2nd, Mazariegos, G.V., Lopez, R., Molmenti, E., Gerber, D., Riddervold, F., Khanna, A., Yin, W.Y., Chen, Y., Scott, V.L., Aggarwal, S., Kramer, D.J., Wagner, R.A., Zhu, Y., Fulmer, M.L., Block, G.D., & Amiot, B.P. (1999). Novel bioartificial liver support system: preclinical evaluation, *Annals of the New York Academy of Sciences*, *875*, 340–352.

[159] Qian, Y., Lanjuan, L., Jianrong, H., Jun, L., Hongcui, C., Suzhen, F., Xia, Y., & Shuhong, Y. (2003). Study of severe hepatitis treated with a hybrid artificial liver support system, *International Journal of Artificial Organs*, *26*, 507–513.

[160] Morsiani, E., Brogli, M., Galavotti, D., Bellini, T., Ricci, D., Pazzi, P., & Puviani, A.C. (2001). Long-term expression of highly differentiated functions by isolated porcine hepatocytes perfused in a radial-flow bioreactor, *Artificial Organs*, *25*, 740–748.

[161] David, B., Dore, E., Jaffrin, M.Y., & Legallais, C. (2004). Mass transfers in a fluidized bed bioreactor using alginate beads for a future bioartificial liver, *International Journal of Artificial Organs*, *27*, 284–293.

[162] Bazou, D., Coakley, W.T., Hayes, A.J., & Jackson, S.K. (2008). Long-term viability and proliferation of alginate-encapsulated 3-D HepG2 aggregates formed in an ultrasound trap, *Toxicology In Vitro*, *22*, 1321-1331.

[163] Fonsato, V., Herrera, M.B., Buttiglieri, S., Gatti, S., Camussi, G., & Tetta, C. (2010). Use of a rotary bioartificial liver in the differentiation of human liver stem cell, *Tissue Engineering Part C, Methods*, *16*, 123-132.

[164] Pavlic, M., Libiseller, K., Hermann, M., Hengster, P., Margreiter, R., & Wurmm, M. (2007). Small human hepatocytes in rotary culture for treatment of alcohol addicts? A pilot study, *Alcoholism Clinical & Experimental Research*, *31*, 729-736.

[165] Wurm, M., Lubei, V., Caronna, M., Hermann, M., Buttiglieri, S., Bodamer, O., Muehl, A., Tetta, C., Margreiter, R., & Hengster, R. (2009). Introduction of a novel prototype bioartificial liver support system utilizing small human hepatocytes in rotary culture, *Tissue Engineering Part A*, *15*, 1063-1073.

[166] Wurm, M., Woess, C., K. Libiseller, Beer, B., & Pavlic, M. (2010). Challenging small human hepatocytes with opiates: further characterization of a novel prototype bioartificial liver, *Tissue Engineering Part A*, *16*, 807-813.

[167] Discher, D.E., Mooney, D.J., & Zandstra, P.W. (2009). Growth factors, matrices, and forces combine and control stem cells, *Science*, *324:* 1673–1677.

[168] Engler, A.J., Sen, S., Sweeney, H.L., & Discher, D.E. (2006). Matrix elasticity directs stem cell lineage specification, *Cell*, *126*, 677–689.

[169] Lee, J., Cuddihy, M.J., Kotov, N.A. (2008). Three-dimensional cell culture matrices: state of the art, *Tissue Engineering Part B*, *14*, 61–86.

[170] Amenta, P.S., & Harrison, D. (1997). Expression and potential role of the extracellular matrix in hepatic ontogenesis: a review, *Microscopy Research and Technique*, *39*, 372–386.

[171] McClelland, R., Wauthier, E., Uronis, U., & Reid, L. (2008). Gradients in the liver's extracellular matrix chemistry from periportal to pericentral zones: influence on human hepatic progenitors, *Tissue Engineering Part A*, *14*, 59–70.

[172] Abu-Absi, S.F., Friend, J.R., Hansen, L.K., & Hu, W.-S. (2002). Structural polarity and functional bile canaliculi in rat hepatocyte spheroids, *Experimental Cell Research*, *274*, 56–67.

[173] Ohno, M., Motojima, K., Okano, T., & Taniguchi, A. (2009). Maturation of the extracellular matrix and cell adhesion molecules in layered co-cultures of HepG2 and endothelial cells, *Journal of Biochemistry*, *145*, 591–597.

[174] Ng, S., Wu, Y.-H., & Zhou, Y. (2005). Optimization of 3-D hepatocyte culture by controlling the physical and chemical properties of the extra-cellular matrices, *Biomaterials*, *16*, 3153–3163.

[175] Elkayam, T., Amitay-Shaprut, S., Dvir-Ginzberg, M., Harel, T., & Cohen, S. (2006). Enhancing the drug metabolism activities of C3A—a human hepatocyte cell line—by tissue engineering within alginate scaffolds, *Tissue Engineering*, *12*, 1357–1368.

[176] Chang, T.T., & Hughes-Fulford, M. (2009). Monolayer and spheroid culture of human liver hepatocellular carcinoma cell line cells demonstrate distinct global gene expression patterns and functional phenotypes, *Tissue Engineering Part A*, *15*, 559–567.

[177] Koyama, T., Ehashi, T., Ohshima, N., & Miyoshi, H. (2009). Efficient proliferation and maturation of fetal liver cells in three-dimensional culture by stimulation of oncostatin M, epidermal growth factor, and dimethyl sulfoxide, *Tissue Engineering Part A*, *15*,1099–1107.

[178] Olsen, B.R. (1996). Role of cartilage collagens in formation of the skeleton, *Annals of the New York Academy of Sciences*, *785*, 124-130.

[179] Imamura, T., Cui, L., & Teng, R. (2004). Embryonic stem cell-derived embryoid bodies in three-dimensional culture system form hepatocyte-like cells in vitro and in vivo, *Tissue Engineering*, *10*, 1716–1724.

[180] Turner, W.S., Schmelzer, E., McClelland, R., Wauthier, E., Chen, W., & Reid, L.M. (2007). Human hepatoblast phenotype maintained by hyaluronan hydrogels, *Journal of Biomedical Materials Research Part B: Applied Biomaterials*, *82*, 156-168.

[181] Bettinger, C.J., Cyr, K.M., Matsumoto, A., Langer, R., Borenstein, J.R., & Kaplan, D.L. (2007). Silk fibroin microfluidic devices, Advanced Materials, *19*, 2847–2850.

[182] Elcin, Y.M., Dixit, V., & Gitnick, G. (1998). Hepatocyte attachment on biodegradable modified chitosan membranes: in vitro evaluation for the development of liver organoids, Artificial Organs, *22*, 837–846.

[183] Feng, Z.Q., Chu, X., Huang, N.P., Wang, T., Wang, Y., Shi, X., Ding, Y., & Gu, Z.Z. (2009). The effect of nanofibrous galactosylated chitosan scaffolds on the formation of rat primary hepatocyte aggregates and the maintenance of liver function, *Biomaterials*, *30*, 2753-2763.

[184] Ho, Y.C., Mi, F.L., Sung, H.W., & Kuo, P.L. (2009). Heparin-functionalized chitosan-alginate scaffolds for controlled release of growth factor, *International Journal of Pharmaceutics*, *376*: 69-75.

[185] Dawson, D., Mapili, G., Erickson, R., Taqvi, S., & Roy, K. (2008) Biomaterials for stem cell differentiation, *Advanced Drug Delivery Reviews*, *60*, 215–228.

[186] Stevens, M.M., Qanadilo, H.F., Langer, R., & Shastri, V.P. (2004) A rapid-curing alginate gel system: utility in periosteum-derived cartilage tissue engineering, *Biomaterials*, *25*, 887–894.

[187] Augst, A.D., Kong, H.J., & Mooney, D.J. (2006). Alginate hydrogels as biomaterials, *Macromolecular Bioscience*, *6*, 623–633.

[188] Fang, S., Qiu, Y.D., Mao, L., Shi, X.L., Yu, D.C., & Ding, Y.T. (2007). Differentiation of embryoid-body cells derived from embryonic stem cells into hepatocytes in alginate microbeads in vitro, *Acta Pharmacologica Sinica*, *28*, 1924-1930.

[189] Cheng, N., Wauthier, E., & Reid, L.M. (2008). Mature human hepatocytes from ex vivo differentiation of alginate-encapsulated hepatoblasts, *Tissue Engineering Part* A, *14*, 1–7.

[190] Liu, C., Xia, Z., & Czernuszka, J.T. (2007). Design and development of three-dimensional scaffolds for tissue engineering, *Chemical Engineering Research and Design*, *85*, 1051–1064.

[191] Lü, J.M., Wang, X., Marin-Muller, C., Wang, H., Lin, P.H., Yao, Q., & Chen, C. (2009). Current advances in research and clinical applications of PLGA-based nanotechnology, *Expert Reviews in Molecular Diagnostics*, *9*, 325-341.

[192] Huang, H., Hanada, S., Kojima, N., & Sakai, N. (2006). Enhanced functional maturation of fetal porcine hepatocytes in three-dimensional poly-L-lactic acid scaffolds: a culture condition suitable for engineered liver tissues in large-scale animal studies, *Cell Transplantation*, *15*, 799–809.

[193] Jiang, J.L., Kojima, N., Guo, L., Naruse, K., Makuuchi, M., Miyajima, A., Yan, W., & Sakai, Y. (2004). Efficacy of engineered liver tissue based on poly-L-lactic acid scaffolds and fetal mouse liver cells cultured with oncostatin M, nicotinamide, and dimethyl sulfoxide, *Tissue Engineering*, *10*, 1577–1586.

[194] Hanada, S., Kojima, N., & Sakai, Y. (2008). Soluble factor dependent in vitro growth and maturation of rat fetal liver cells in a three-dimensional culture system, *Tissue Engineering Part A*, *14*, 149–160.

[195] Sokolsky-Papkov, M., Agashi, K., Olaye, A., Shakesheff, K., & Domb, A.J. (2007). Polymer carriers for drug delivery in tissue engineering, *Advanced Drug Delivery Reviews*, *59*, 187–206.

[196] Desai, J.P., Pillarisetti, A., & Brooks, A.D. (2007). Engineering approaches to biomanipulation, *Annual Reviews in Biomedical Engineering*, *9*, 35–53.

[197] Yamashita, Y., Shimada, M., Tsujita, E., Shirabe, K., Ijima, H., Nakazawa, K., Sakiyama, R., Fukuda, J., Funatsu, K., & Sugimachi, K. (2002). High metabolic function of primary human and porcine hepatocytes in a polyurethane foam/spheroid culture system in plasma from patients with fulminant hepatic failure, *Cell Transplantation,* 11, 379-384.

[198] Fukuda, J., Okamura, K., Ishihara, K., Mizumoto, H., Nakazawa, K., Ijima, H., Kajiwara, T., & Funatsu, K. (2005)._Differentiation effects by the combination of spheroid formation and sodium butyrate treatment in human hepatoblastoma cell line (Hep G2): a possible cell source for hybrid artificial liver, *Cell Transplantation, 14,* 819-827.

[199] Yamashita, Y., Shimada, M., Tsujita, E., Shirabe, K., Ijima, H., Nakazawa, K., Sakiyama, R., Fukuda, J., Funatsu, K., & Sugimachi, K. (2003). Efficacy of a larger version of the hybrid artificial liver support system using a polyurethane foam/spheroid packed-bed module in a warm ischemic liver failure pig model for preclinical experiments, *Cell Transplantation*, 12, 101-107.

[200] Shimada, M., Yamashita, Y., Tanaka, S., Shirabe, K., Nakazawa, K., Ijima, H., Sakiyama, R., Fukuda, J., Funatsu, K., & Sugimachi, K. (2007). Characteristic gene expression induced by polyurethane foam/spheroid culture of hepatoma cell line, Hep G2 as a promising cell source for bioartificial liver, Hepatogastroenterology, *54*, 814-820.

[201] Matsumoto, K., Mizumoto, H., Nakazawa, K., Ijima, H., Funatsu, K., & Kajiwara, T. (2008). Hepatic differentiation of mouse embryonic stem cells in a three-dimensional culture system using polyurethane foam, *Journal of Bioscience and Bioengineering*, *105*, 350–354.

[202] Hou, Y.T., Ijima, H., Takei, T., Kawakami, K. (in press). Growth factor/heparin-immobilized collagen gel system enhances viability of transplanted hepatocytes and induces angiogenesis, Journal of Bioscience and Bioengineering, doi: 10.1016/j.jbiosc.2011.05.003.

[203] Loffredo, F., & Lee, R.T. (2008). Therapeutic vasculogenesis: it takes two, *Circulation Research*, *103*, 128–130.

[204] Poyck, P.P.C., Mareels, G., Hoekstra, R., van Wijk, A.C., van der Hoeven, T.V., van Gulik, T.M., Verdonck, P.R., & Chamuleau, R.A. (2008). Enhanced oxygen availability improves liver-specific functions of the AMC bioartificial liver, *Artificial Organs*, *32*, 116–126.

[205] Oh, S.H., Ward, C.L., Atala, A., Yoo, J.J., & Harrison, B.S. (2009). Oxygen generating scaffolds for enhancing engineered tissue survival, *Biomaterials*, *30,* 757–762.

[206] Nahmias, Y., Schwartz, R.E., Hu, W.S., Verfaillie, C.M., & Odde, D.J. (2006). Endothelium-mediated hepatocyte recruitment in the establishment of liver-like tissue in vitro, *Tissue Engineering*, *12*, 1627–1638.

[207] Stolz, D.B., Ross, M.A., Salem, H.M., Mars,W.M., Michalopoulos, G.K., & Enomoto, K. (1999). Cationic colloidal silica membrane perturbation as a means of examining changes at the sinusoidal surface during liver regeneration, *American Journal of Pathology*, *155*, 1487–1498.

[208] Auth, M.K., Joplin, R.E., Okamoto, M., Ishida, Y., McMaster, P., Neuberger, J.M., Blaheta, R.A., Voit, T., & Strain, A.J. (2001). Morphogenesis of primary human biliary epithelial cells: induction in high-density culture or by coculture with autologous human hepatocytes, *Hepatology*, *33*, 519-529.

[209] Auth, M.K., Woitaschek, D., Beste, M., Schreiter, T., Kim, H.S., Oppermann, E., Joplin, R.E., Baumann, U., Hilgard, P., Nadalin, S., Markus, B.H., & Blaheta, R.A. (2005). Preservation of the synthetic and metabolic capacity of isolated human hepatocytes by coculture with human biliary epithelial cells, *Liver Transplantation*, *11*, 410-419.

[210] Soto-Gutierrez, A., Navarro-Alvarez, N., Yagi, H., Nahmias, Y., Yarmush, M.L., & Kobayashi, N. (2010). Engineering of an hepatic organoid to develop liver assist devices, *Cell Transplantation*, *19*, 815-822.

[211] Linke, K., Schanz, J., Hansmann, J., Walles, T., Brunner, H., & Mertsching, H.(2007). Engineered liver-like tissue on a capillarized matrix for applied research, *Tissue Engineering*, *13*, 2699-2707.

[212] Saito, R., Ishii, Y., Ito, R., Nagatsuma, K., Tanaka, K., Saito, M., Maehashi, H., Nomoto, H., Ohkawa, K., Mano, H., Aizawa, M., Hano, H., Yanaga, K., & Matsuura, T. (2011). Transplantation of liver organoids in the omentum and kidney, *Artificial Organs*, *35*, 80-83.

[213] Uygun, B.E., Price, G., Saedi, N., Izamis, M.L., Berendsen, T., Yarmush, M., Uygun, K. (2011). Decellularization and recellularization of whole livers, *Journal of Visualized Experiments*, *(48)*: pii: 2394.

[214] Baptista, P.M., Siddiqui, M.M., Lozier, G., Rodriguez, S.R., Atala, A., & Soker, S. (2011). The use of whole organ decellularization for the generation of a vascularized liver organoid, *Hepatology*, *53*: 604-617.

[215] Uygun, B.E., Soto-Gutierrez, A., Yagi, H., Izamis, M.L., Guzzardi, M.A, Shulman, C., Milwid, J., Kobayashi, N., Tilles, A., Berthiaume, F., Hertl, M., Nahmias, Y., Yarmush, M.L., & Uygun, K. (2010). Organ reengineering through development of a transplantable recellularized liver graft using decellularized liver matrix, *Nature Medicine*, *16*, 814-820.

[216] Crapo, P.M., Gilbert, T.W., & Badylak, S.F. (2011). An overview of tissue and whole organ decellularization processes, *Biomaterials*, *32*, 3233-3243.

[217] Macchiarini, P., Jungebluth, P., Go, T., Asnaghi, M.A., Rees, L.E., Cogan, T.A., Dodson, A., Martorell, J., Bellini, S., Parnigotto, P.P., Dickinson, S.C., Hollander, A.P., Mantero, S., Conconi, M.T., & Birchall, M.A. (2008). Clinical transplantation of a tissue-engineered airway, *Lancet*, *372*, 2023-2030.

In: Tissue Engineering
Editors: T. Ishikawa and M. Yamamoto
ISBN 978-1-62257-062-1

Chapter 2

CARBON NANOTUBE CONTAINING BIOCERAMIC AND BIOPOLYMER COMPOSITE SCAFFOLDS FOR BONE TISSUE ENGINEERING

***Jayachandran Venkatesan*[1,2] *and Se Kwon Kim*[1,2*]**
[1]Department of Chemistry, Pukyong National University, Busan, Republic of Korea
[2]Marine Bioprocess Research Center, Pukyong National University, Busan, Republic of Korea

ABSTRACT

Tissue engineering is a vast and emerging area of research in the recent years. It has been significantly reduced the secondary operation and transmissible disease as well as the huge amount of expensive compared with allograft and allograft techniques. Bone is made of hydroxyapatite as well as collagen as a major portion. The great interest has been given by the researchers in the preparation of artificial bone like materials.

For this, scaffolds are made by natural based polymers cellulose, chitin, chitosan, alginic acid and amylopectin are commonly employed. The addition of bioceramics (nano hydroxyapatite and tricalcium phosphate) and other promising materials (carbon nanotube, grapheme and boron nitride nanotubes) with biopolymers are great interest to accomplish the goal in the preparation of bone like material. In this chapter, the preparation, characterization and *in vitro* biological activity of double and triple component system scaffolds have been explored for bone tissue engineering application.

Keywords: Chitosan; Hydroxyapatite; Carbon nanotube; Graphene; Boron nitride nanotube and Bone tissue engineering

* Corresponding author Tel: +82 51 629 7097; Fax: +82 51 628 8147. E-mail address: sknkim@pknu.ac.kr.

1. Introduction

Research on biomaterials for bone implantation and replacement has expanded considerably over the last four decades. In recent years, significant progress has been made in organ transplantation, surgical reconstruction and the use of artificial protheses to treat the loss or failure of an organ or bone tissue.

The establishment of a load bearing biomaterial must be incorporated with natural bone. The implanted biomaterial should possess the following criteria: biocompatibility, osteoconductivity, high porosity and biomechanical compatibility [1].

For this requirement, autografts and allografts are used extensively for bone grafts. In the autograft technique, bone from another part will be harvested within the body, and this material fills the gap and provides optimal osteoinductivity, osteoconductivity and osteogenic properties. However, it has its own disadvantages: autografting often leads to complications in wound healing, additional surgery, donor pain and an inadequate supply of bone to fill the gap [2].

In the allograft technique, cadaver bones have been used, but it has problems with immunogenic reactions and the risk of acquiring transmissible diseases (AIDS and hepatitis) from tissues and fluids. These limitations and concerns have created substantial interest in the development of artificial materials as bone graft substitutes [3].

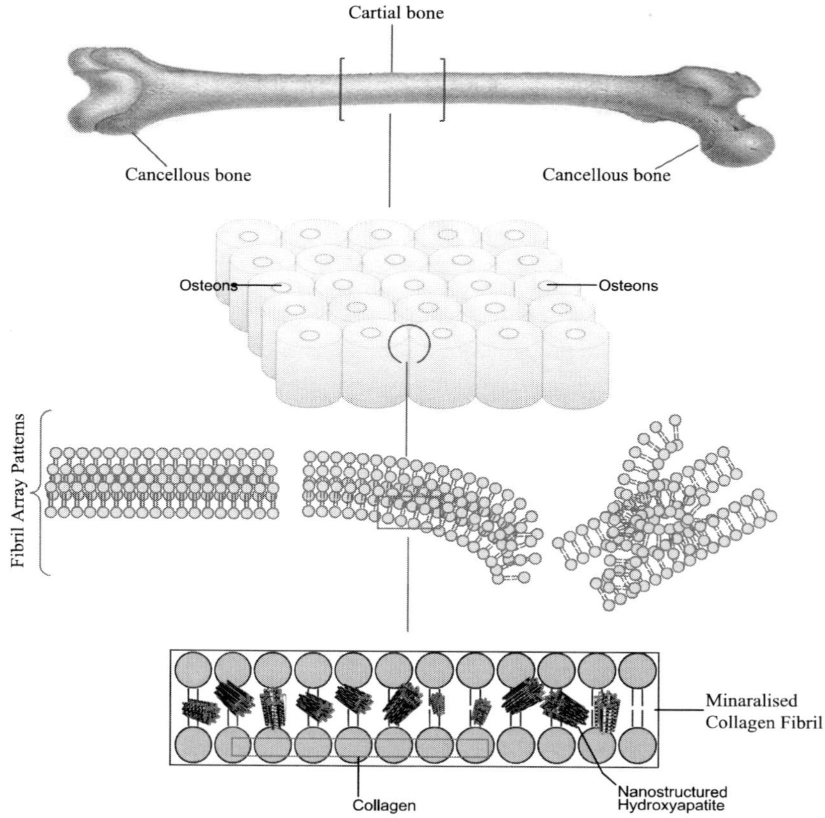

Figure 1. Hierarchical levels organization of bone.

Very few compounds are classified as bioactive, biodegradable and osteoconductive. Biopolymers and bioceramics are the best bioactive biomaterials in bone tissue engineering and renowned for their excellent biocompatibility with the human body environment [4].

2. Constituent of Bone and Structure

Bone is a complex living tissue with hierarchical structure that consists of 70% calcium phosphate mineral (mostly nano scale hydroxyapatite crystals) and 30% organic moieties (including collagen, glycoproteins, proteoglycans, and sialoproteins) by dry weight as shown in Figure 1 [5-9, 11].

There are four kinds of cells which are playing most important function for bone remodeling, they are:

- Osteoblasts - Bone formation
- Osteoclasts - Bone resorption
- Osteocytes - A mature osteoblast which no longer secretes matrix.
- Osteoprogenitor - Immature cells which differentiate to make osteoblasts

Osteoblast and osteoclast cells are responsible for bone formation and resorption, respectively. Calcium and phosphate are the two minerals that are essential for normal bone formation. Osteoblasts secrete a calcifiable matrix which contains minerals, collagen and also small amount of non collagenous proteins including osteopontin, osteonectin, bone sialoprotein, and osteocalcin [10].

3. Tissue Engineering

Today, tissue engineering is considered to be one of the most prominent and important modus operandi for various biomedical applications. It is categorized as a sub-field of bio materials and use of combination of cells, engineering, materials and medical science. While most definitions of tissue engineering cover a broad range of applications, in practice the term is closely associated with applications that repair or replace portions of or whole tissues (i.e., bone, cartilage, blood vessels, bladder, skin and etc.).

The potential impacts of tissue engineering are,

- Reduces the need for organ replacement,
- Greatly accelerates the development of new drugs that may cure patients,
- Eliminates the high expenses for artificial materials.

The loss or failure of an organ or tissue is one of the most frequent, devastating and costly problem in human health care [12].

4. Biomaterials for Bone Tissue Engineering

Biomaterials are natural or synthetic nonviable materials, which in whole or in part composed of living matrix or composites such as polymer scaffolding that is suitable for introduction into living tissue, especially as part of a medical device [13]. Any material that is intended to interface with biological systems to evaluate, treat, augment, or replace any tissue in the body can be a biomaterial. Because of their incredible applications in tissue engineering, organ transplantation and other biomedical studies, the development strategies for these materials are increasing day by day [14]. However, the development of biomaterials is not a new area of science, which existed for around half a century [15]. Various materials have been utilized to produce scaffolds for several applications. Ceramic materials are also used particularly in combination with polymers especially in bone tissue applications, thus forming composite materials with improved mechanical and biological properties. The requirements for a scaffold in bone regeneration are:

- Biocompatibility,
- Osteoconductivity,
- Interconnected porous structure,
- Appropriate mechanical strength and
- Biodegradability.

Several methods have been used to design 3D scaffolds for bone tissue engineering such as freeze with lyophilization, solvent-casting and particulate leaching technique, gas foaming process, emulsion freeze drying, electro spinning technique, rapid-prototyping techniques as well as thermally induced phase separation [16].

The lists of commercially available bone filling materials have shown in the Figure 2. Moreover, ceramic compounds alone cannot mimic all the functions of a natural bone. Thus, combining it with a polymer has great potential value to mimic the natural functions. There are two kinds of polymers viz. synthetic and natural polymers. Synthetic polymers are often used in bone tissue engineering. Biodegradable polymers which are chemically synthesized are widely used as scaffold materials in tissue engineering because they typically offer high versatility, stable properties and good workability. Therefore tailoring synthetic polymers is easier as compared to natural polymers and a wider range of shape and properties can be obtained; furthermore the final result is more predictable and the scaffolds are more uniform.

Degradation rate of scaffolds can be adapted to the specific applications by selecting specific polymers, copolymers or blends. Most of these polymers undergo to a simple hydrolytic degradation. However, biocompatibility of synthetic polymers is generally lower than natural polymers.

Among these bioresorbable polymers, poly glycolic acid, poly lactic acid or their copolymers or blends as well as the aliphatic polyester poly caprolactone are mostly used as scaffolds. But their biocompatibility and biodegradability is limited when compared to natural polymers. In the light of above mentioned views, this research focuses on the use of natural polymers in bone tissue engineering. Naturally derived polymer materials such as chitosan, collagen, gelatin, chondroitin sulfate and amylopectin can be extracted from plants, animals or human tissues; they exhibit good biocompatibility, inexpensive, low toxicity and a low chronic inflammatory response. They can be combined into a composite with other natural materials or synthetic materials and can be degraded by naturally occurring enzymes.

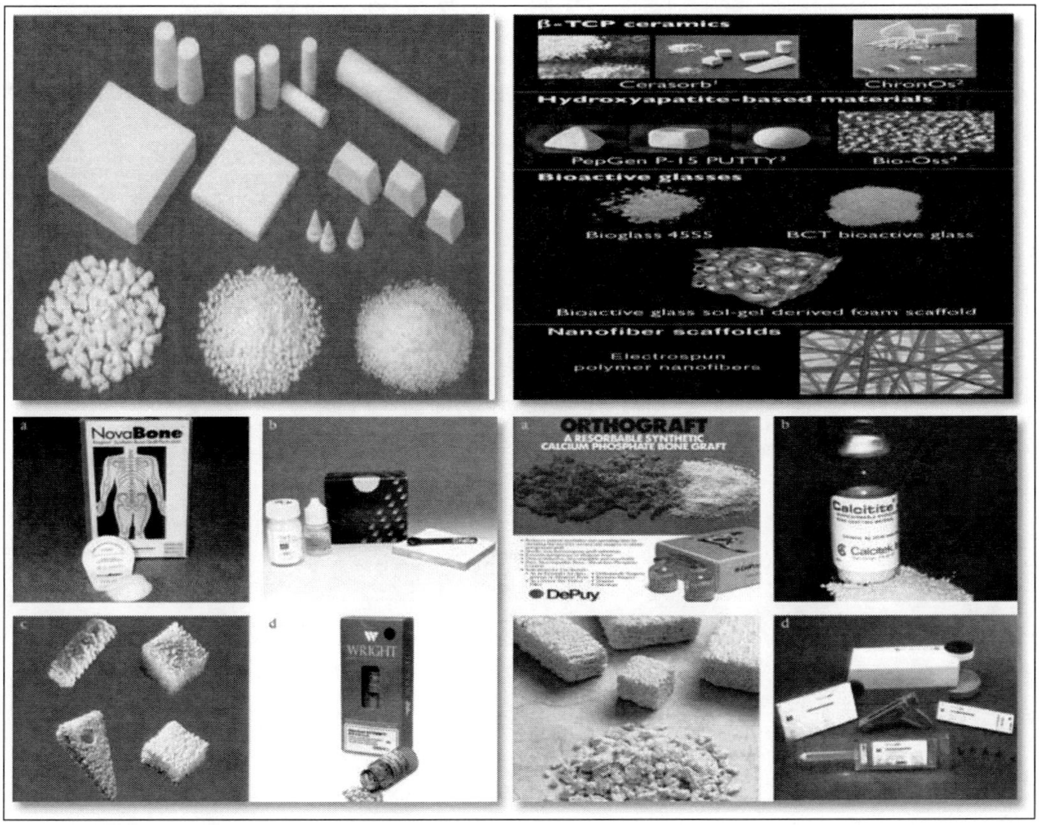

[Figure adopted from Reference no.[17]].

Figure 2. Commercially available ceramic scaffolds for bone tissue engineering.

4.1. Biopolymers

Chitosan (Figure 3) is the N-deacetylated derivative of chitin, a natural polysaccharide commonly located in the exoskeleton of crustaceans and insects. It is a linear polysaccharide composed of β (1-4) linked D-glucosamine with randomly dispersed N-acetyl-D-glucosamine groups. Chitosan can undergo degradation by enzymes such as chitosanase and lysozyme. There is a big interest for this natural polymer in tissue engineering field because it is a well tolerant implant material, non-antigenic, biodegradable and it has good structural properties. Considerable attention has been given to chitosan and its derivative materials and their applications in the field of orthopedic tissue engineering. Interesting characteristics that render chitosan suitable for this purpose are a minimal foreign body reaction, an intrinsic antibacterial nature and the ability to be molded in various geometries and forms such as porous structures and osteoconduction. Due to its favorable gelling properties chitosan can deliver morphogenic factors and pharmaceutical agents in a controlled fashion [18].

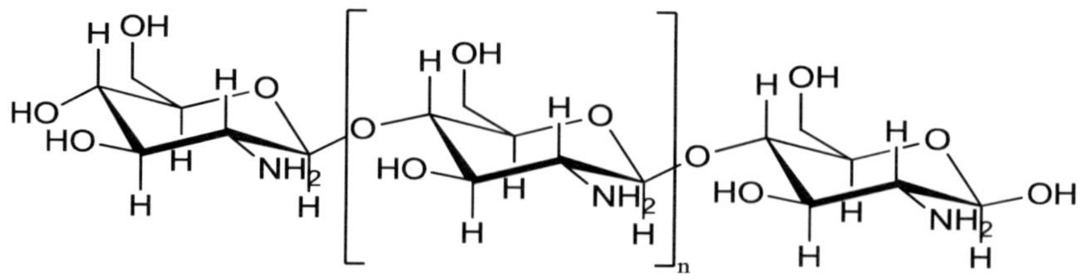

Figure 3. Structure of chitosan.

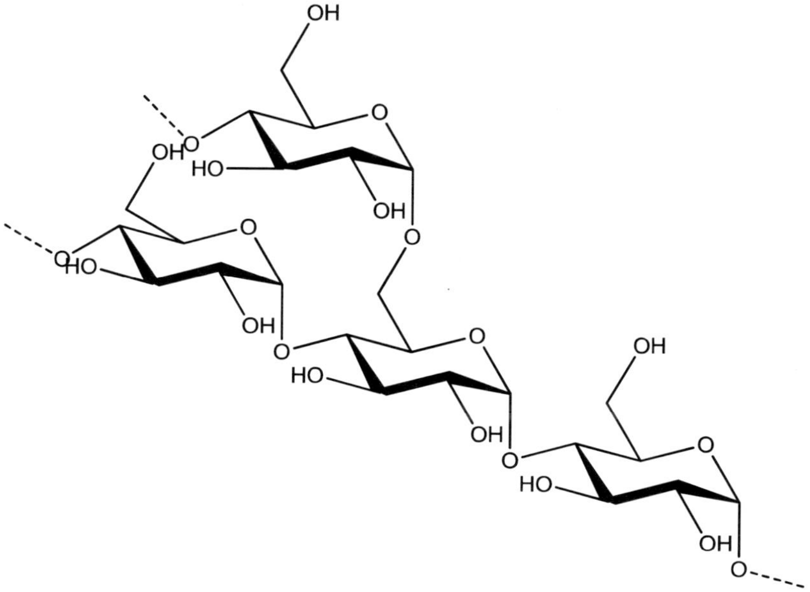

Figure 4. Structure of Amylopectin.

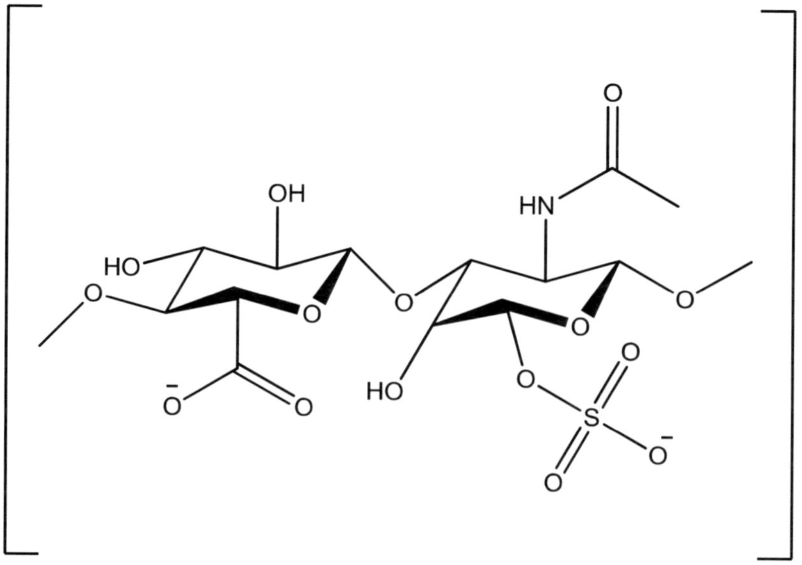

Figure 5. Structure of Chondroitin sulfate.

Amylopectin and Chondroitin sulfate are branched and sulfated polysaccharides and are widely studied for bone tissue engineering (Figure 4 and 5).

4.2. Bioceramics

There have been a number of major advances in the field of bioactive ceramics, glasses and glass ceramics during the past 30–40 years [19]. Biodegradable polymers and bioactive ceramics are being combined in a variety of composite materials for tissue engineering scaffolds. Materials and fabrication routes for 3D scaffolds with interconnected high porosities suitable for bone tissue engineering [20-22]. Calcium phosphate minerals, hydroxyapatite Ca_{10} $(PO_4)_6(OH)_2$ (HAp) is considered to play a vital role in various fields including spinal fusion, craniomaxillofacial reconstruction, bone defects, fracture treatment, total joint replacement (bone augmentation), revision surgery [19, 23], catalysis, fertilizers, pharmaceutical products, protein chromatography, water treatment process and preparation of artificial bone [24]. HAp is the most stable calcium phosphate salt at normal temperatures and pH between 4 and 12. HAp derived from powder processing method has a great potential as bone substitute owing to its excellent biocompatible and osteoconductive properties. Several methods have been reported for the synthesis of HAp crystals in the literature, hydrothermal [25], liquid membrane [26], precipitation [27], radio frequency thermal plasma [28], ultrasonic precipitation [29], reverse micro emulsion [30], sol-gel [31] and polymer assisted method [32-33] are some of them.

5. Stem Cell Therapy with Biomaterials

Stem cells play increasingly prominent roles in tissue engineering and regenerative medicine (Figure 6). But the limitation in the translation of stem cell technologies to clinical applications is the limited supply of cells. Advances in biomaterial engineering and scaffold fabrication enable the development of ex vivo cell expansion systems that are being widely used these days.

Progress in biomaterial design has also allowed directed differentiation of stem cells into specific lineages. In addition to delivering biochemical cues, various technologies have been developed to introduce micro and nano scale features onto culture surfaces to enable the study of stem cell responses to topographical cues [34].

Based on the concept that each tissue is characterized by different functions and features; the tissue engineering product design strongly depends on material properties and on cellular response to the chosen biomaterials. In addition, the ability to shape into the suitable cellular level architecture has to be taken into account and the final architecture must be compatible with the desired tissue response. The proper biomaterial selection is assisted by the development of methods and sophisticated modeling techniques that permit prediction of polymer properties and cellular response to the material. Such techniques allow a wide use of biodegradable polymers for tissue engineering applications. Tissue engineering products can be designed to conduct, induct or block tissue responses and architectures.

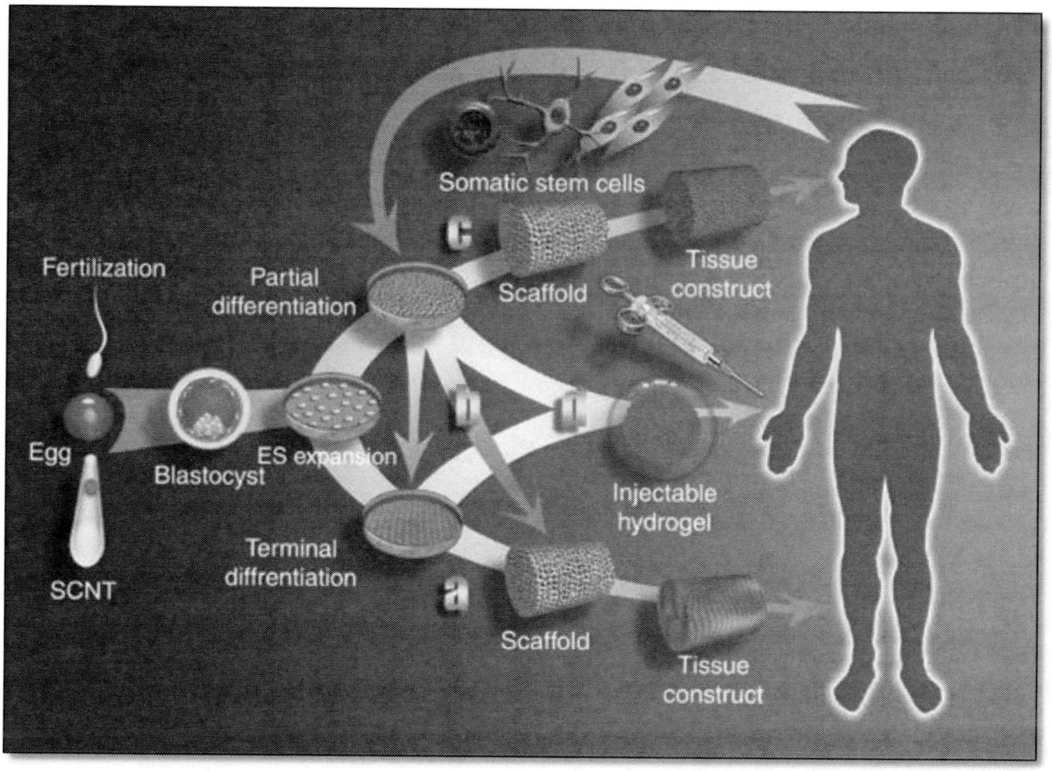

[Figure adopted from Ref. [34]].

Figure 6. Role of biomaterials in stem cells therapy.

The term regenerative medicine is often used synonymously with tissue engineering, although those involved in regenerative medicine place more emphasis on the use of stem cells to produce tissues. Generally, tissue engineering revolves around the use of living cell and/or extracellular matrix components in the development of implantable parts or devices and in particular one of the main targets of tissue engineering concerns cell seeding on porous biodegradable three dimensional (3D) scaffolds, followed by implantation of the scaffold into the injured area to provide regeneration.

6. Scaffolds in Bone Tissue Engineering

Tissue engineering scaffolds or matrices are an important component for bone tissue development and their characteristics such as pore size and the structure, ordered and regular or randomly distributed, are fundamental in forming tissues with proper cell morphology, orientation, arrangement of intercellular material, and the relationship between different cell types. Besides several techniques have been developed to form well defined scaffolds using different kind of biomaterials and physical characteristics of the matrices can be varied to maximize cellular and tissue responses [35].

The development of 3D scaffolds, support structures or matrices that induce cells to form functional tissues, is one of the main object of tissue engineering. In particular by using scaffolds, transplanted cells can be delivered to a specific site in a tissue driving the growth of cells inside a desired anatomic site. Thus scaffolds represent the space available for the tissue to develop and act as a physical support for cells growth. 3D porous matrices provide mechanical and adhesive support for seeded cells in culture and tissue in growth allowing vascularisation and shape maintenance during tissue regeneration. Mechanical support against compressive or tensile forces present inside the physiological environmental system of the human body should be taken into account depending on the tissue the scaffold is designed for. Enough mechanical strength and stiffness is needed to initially contrast wound contraction forces and later to guarantee a perfect reconstruction of the tissue. For this reason scaffold degradation profile must be designed such that it can support the tissue formation until new tissue is developed.

Therefore the primary function of a scaffold is tissue conduction and thus it must allow cell attachment, migration onto or within the scaffold, cell proliferation and differentiation. The morphology of the scaffold should be designed to be able to guide the formation of the new tissue in terms of pore size, shape and vascularisation.

Scaffold geometry should also define the space and the outer shape of the defect to finally be properly adapted, also matching the healthy tissue stiffness and strength while maintaining an interconnected pore network for cell migration and nutrient transportation. It must also provide an environment in which cells can maintain their phenotype and synthesize proteins and molecules. Scaffolds compatible to the structure of the tissue should be designed to have high porosity, high surface area, fully interconnected geometry, structural strength and a specific three dimensional shape.

Besides, scaffold materials should be biocompatible and biodegradable or resorbable so as to allow replacement by newly formed tissue in long term. A porous structure is usually needed for two critical functions. First, pore channels provide ports of entry for migrating cells or for capillary suction of blood. Secondly, a large area is available for specific and numerous cell interactions. In particular micro porosity is fundamental for capillary ingrowth and cell matrix interactions while macro porosity for nutrients supply and waste removal of cells grown on the scaffold. It is important to underline that an ideal scaffold design does not exist and each tissue requires a specific matrix designed with defined material properties. Finally scaffolds should be manufactured in a reproducible, controlled and cost effective fashion with the possibility to include biological components, such as cells and grow factors.

Although many of the materials and devices developed for various applications, the core materials are organic matrices are e.g., the polysaccharide derivatives and the asymmetric protein collagen. These are representing as the basic structural modules and organo-templates for biomineralization and biomimicry in nature [36]. Especially, as marine bioprocessed byproducts, chitin derived chitosans and their oligosaccharides are most proven biocompatible and biodegradable natural materials, which show tremendous biological activities and applications [37-39]. On the other hand, collagen, being one of the highly challenging biopolymers similar to chitosans, are applied in various biomimetic scaffold /composite generation, biomineralization etc. [40-44].

Collagens are vastly used as collagen film/sheet/disc, collagen nanoparticles/nanospheres, calcifiable matrix system, collagen gel/hydrogel, collagen-liposomes, collagen shields, collagen sponges, pellet/tablet etc. The specificity and uniqueness of the function of collagens made them to use as individual or in combination with other polymers like chitosans to server as biomaterials [45].

7. Composite Materials for Bone Tissue Engineering

Composite materials are now playing predominant role as scaffolds in bone tissue engineering. Chitosan has numerous advantageous properties for orthopedic applications, as described above and elsewhere [46], which makes it ideal as a bone graft substituent. Chitosan scaffolds are flexible and their mechanical properties are inferior to those of normal bone, as it is unable to support load bearing bone implants. Moreover, chitosan itself is not osteoconductive, although addition of ceramic materials improves its osteoconductivity and mechanical strength.

Chitosan scaffolds alone cannot imitate all the properties of natural bone. The substantial development of composite materials with chitosan mimics all the properties of bone. As proven, calcium phosphate materials are osteoconductive to mimic the inorganic portion of natural bone, while chitosan/HAp composite materials show promise in mimicking the organic portion as well as the inorganic portion of natural bone. Several studies have been conducted with chitosan /HAp composite materials for bone tissue engineering [46-57]. Calcium phosphate compounds are of great interest in the field of bone tissue engineering. HAp is one of the most stable forms of calcium phosphate and it occurs in the bone as a major component (60 to 65%) [58]. HAp also possesses a variety of uses, including orthopedic, dental and maxillofacial applications. Therefore, HAp has recently emerged as an important compound for artificial bone preparation. It stimulates osteoconduction being gradually replaced by the host bone after implantation. It is being used for orthopedic replacements, especially in bone regeneration and dental implant treatment. The mechanical properties of HAp are poor, though, so it cannot be used for load bearing bone tissues. Polymers have been used to improve the mechanical properties of HAp (compressive strength, Young's modulus, fracture toughness) [59]. When chitosan is combined with HAp, it might be able to mimic the function of natural bone.

8. Carbon Nano Tubes Composite Biomaterials for Bone Tissue Engineering

Carbon nanotube (CNT) is promising biomaterials for tissue engineering in the last two decays. After exordinary development of carbon nano fibers, the recent result turns to knowing the function of those materials [60]. After the discovery of carbon nanotube [61], there is a tremendous research in the field of tissue engineering due to its excellent properties such as electrical and mechanical properties. The researches on CNT were increasing for in the past decades in tissue engineering as expressed in Figure 7.

The role of marine natural polymers is enormous because of their *in situ* biocompatible and biodegradable applications *e.g.* chitosan/hyaluronic acid based hydrogels in cartilage tissue engineering [62]. These biodegradable polymers could be useful for various bone related implantations e.g. as large implants like screws, plates etc. [63].

Chitin derived chitosan and collagen in the form of scaffold or composite can influence the capacity of seeded cells to regenerate bone within the defect site by functioning as an extracellular matrix (ECM), thereby providing surface contact and temporary mechanical support for functional cells and maintaining space for bone tissue development [64].

CNTs are allotropes of carbon with a cylindrical nanostructure and constructed with length-to-diameter ratio of up to 28,000,000:1 (Figure 8).

These cylindrical carbon molecules have novel properties, which make them potentially useful in many applications in nanotechnology, electronics, optics and materials sciences. CNTs have a high Young's modulus (1.0–1.8 TPa), high tensile strength (30–200 GPa) and high elongation at break (10–30%). In addition, they have extremely small size (about 1–10 nm in diameter), high aspect ratio (>1,000), high structural and chemical stability, and stiffness, as well as remarkable electrical, thermal, optical and bioactive properties [65-66]. All these properties make CNTs especially promising candidates as reinforcement fillers in the development of nano composites.

It has been observed that the combination of CNT with chitosan leads to an enormous increase in the mechanical strength of the composite [67].

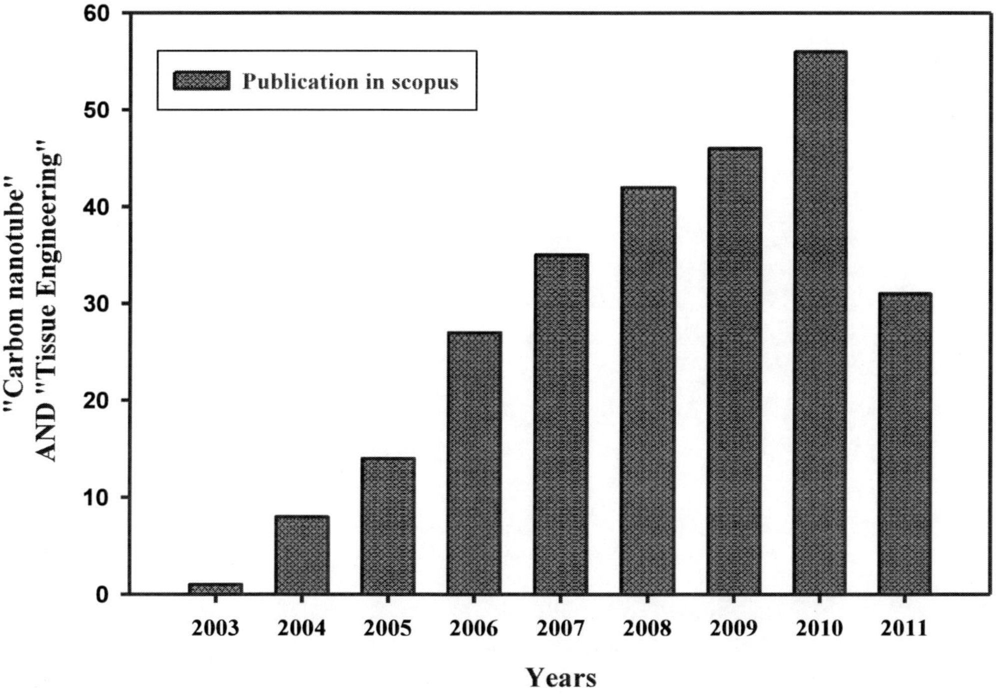

Figure 7. Carbon nanotube research in Tissue engineering (Scopus details).

CNTs hold great interest with respect to biomaterials, particularly those to be positioned in contact with bone such as prostheses for arthroplasty, plates or screws for fracture fixation, drug delivery systems, and scaffolding for bone regeneration. The most important concerns for the use of CNTs as a biomaterial are tissue safety, but only few reports have addressed the toxicity of CNTs. In particular, bone tissue compatibility is extremely important for using CNTs in biomaterials. Usui *et al.*, who first developed CNT, found that these tubes have good bone tissue compatibility and are capable of permitting bone repair and becoming closely integrated with bone tissue and accelerate bone formation stimulated by recombinant human bone morphogenetic protein-2 [68].

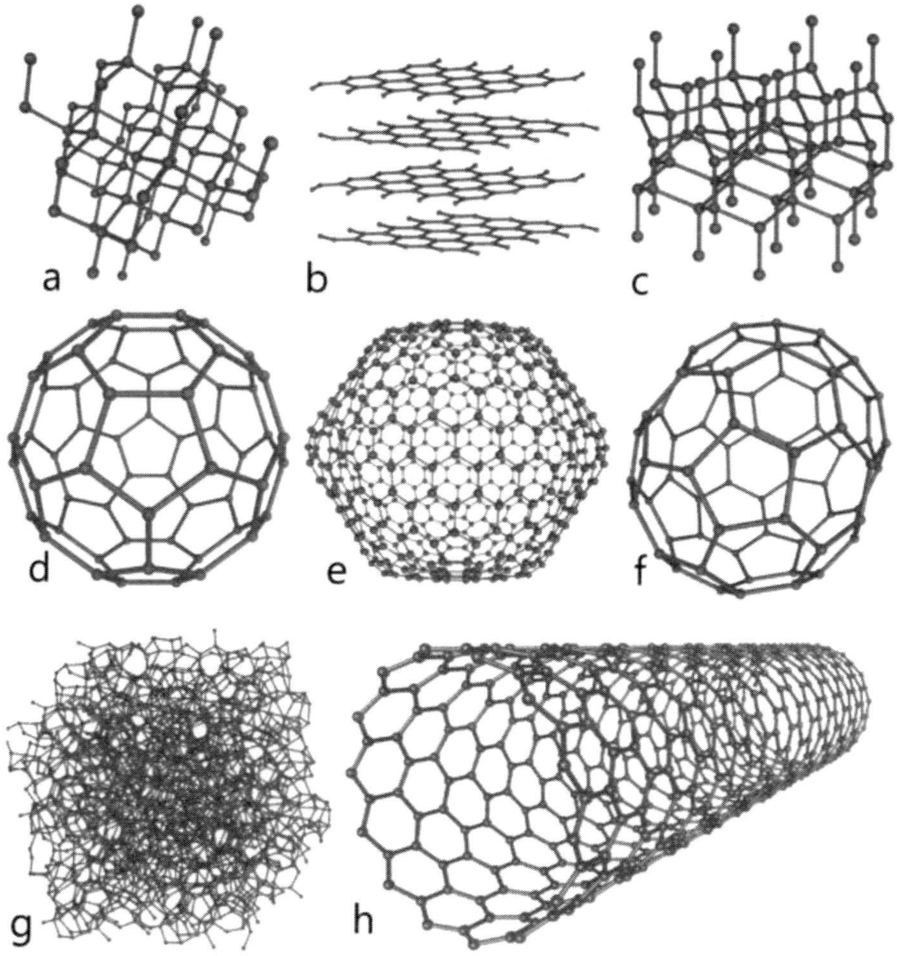

Figure 8. Allotropes of carbon (a) diamond; (b) graphite; (c) lonsdaleite; (d-f) fullerenes (C60, C540 and C70); (g) amorphous carbon and (h) Carbon nanotube.

A big challenge in the bone tissue engineering is mechanical strength improvement of the scaffold materials. One of the main purposes of creating chitosan composites is to improve the mechanical strength of the material. CNT is a promising material to fulfill that gap due to its strong mechanical properties. Several authors have developed many composites as well as scaffold materials with CNT.

It has been observed that cell adhesion on multi-walled CNT (MWCNT) coated dish is much higher than that on the collagen coated dish [69].

8.1. Preparation of Chitosan/CNT Composite

One of the major problems in the preparation of chitosan with CNT materials is the difficulty to disperse the CNTs well in the polymer matrix. Several methods have been used for this purpose such as controlled surface deposition and cross linking process [70], simple solution evaporation method [67], thermally induced phase separation followed by freeze drying [71], electrodeposition [72], spray layer by layer technique [73], pH and electrical actuation [74], wet spinning method [75], electrospinning [76-77] and coagulation and hydrodynamic method [77].

The tricomponent scaffold of chitosan/HAp with carbon nanotube has been prepared via freeze dry method for bone substitute (Figure 9).

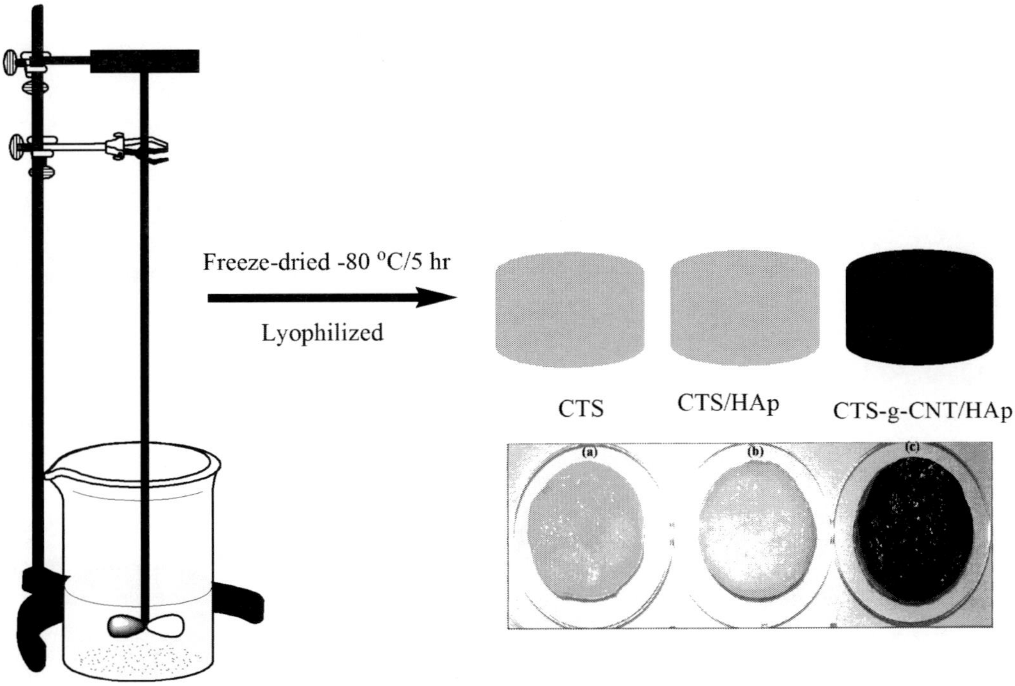

Figure 9. Scaffold preparation with free dry method with chitosan, HAp and carbon nanotube.

8.2. Mechanical Properties of Chitosan/CNT Composite

Wang *et al.* reported that CNT is homogeneously dispersed throughout the chitosan matrix. When 0.8% of CNT was introduced in the chitosan matrix, the mechanical properties, including the tensile modulus and strength of the nanocomposite were greatly improved by about 93% and 99% [67]. A small addition of CNTs significantly improves the tensile properties of chitosan matrix, and the mechanical properties increase with the increase of CNTs' loading (0.4 wt% of CNTs filler) whereas the tensile modulus and strength of the nanocomposite increases dramatically by about 78% and 94%. Due to the aggregation of CNTs within the chitosan matrix at higher concentrations, with further increases of the loading level of CNTs, the tensile modulus only increases slightly, while the tensile strength remains stable. For instance, with 2.0 wt% of MWNTs filler, the tensile modulus of the chitosan matrix is 2.15 GPa, only 0.07 GPa higher than that of 0.8 wt% of MWNTs. The tensile test results indicate the mechanical properties of the chitosan/CNTs [67].

In another study, the tenacity of neat chitosan fiber was recorded as 96 MPa, whereas that of chitosan reinforced with purified but non-functionalized single-walled CNTs (p-SWCNTs) increased from 132 MPa for 0.01 wt% SWCNT addition to 180 MPa for 0.4 wt% SWCNT addition. Even at very low SWCNT concentration (0.01 wt %), a significant increase in tenacity was observed.

On the other hand, on incorporation of functionalized CNT (f-CNT), the tenacity increased from 136 MPa for 0.01 wt% SWCNT addition to 226 MPa for 0.4 wt% additions. This continuous improvement in tenacity (even at higher SWCNT concentrations) may be ascribed to better dispersion as well as interaction of functionalized nanotubes in chitosan [74]. This might be because of the strong chemical interaction as functionalized CNT's COOH group interacts with amine group of chitosan matrix.

8.3. *In Vitro* Study of Chitosan/CNT Composites

CNTs are like an inert matrix material in which cells can proliferate in easily. Zanello *et al.* suggested that CNTs can be used as an alternative material for the treatment of bone pathologies with the potential for the regrowth of normal bone.

Osteoblast-like cells were grown on electrically neutral CNTs and they were also found to produce mineralized bone [78]. The impurities in commercial SWCNTs, such as metal catalysts and amorphous carbon particles, are reported to be toxic to cells and could induce intracellular reactive oxygen species. However, the addition of chitosan with amine functional groups should decrease the cytotoxicity. Compared to p-SWCNTs, f-SWCNT grafted with chitosan provides better cell proliferation and cell attachment. An explanation to this may be that the cellular membranes are negatively charged, and thus cells can more easily attach and grow on more positively charged surfaces [79].

Wörle-Knirsch *et al.* suggested that the MTT method is not reliable for finding out the cytotoxity of CNT based materials, as the MTT formazan crystals formed in the MTT reaction are lumped. As a consequence, false results may show a strong cytotoxic effect in the MTT assay after 24 h and show that roughly 50% of the cells die. But other methods, such as WST-1, lactate dehydrogenase, *fluorescence-activated cell sorting* assisted mitochondrial membrane potential determination and annexin-V/PI staining revealed no cytotoxicity [80].

Very little work has been done with chitosan/CNT composite materials for bone tissue engineering. CNT can mimic the strength of natural bone in composite or scaffold materials. Research strategies still need to be developed in the field of chitosan/CNT composite material preparation and *in vitro* as well as *in vivo* activity of CNT based polymeric materials.

8.4. Chemical Interactions between Chitosan/CNT

The raw CNTs are mainly hydrophobic and poorly miscible with water. The acid treated CNTs contain many defects and hydrophilic groups, such as -OH and -COOH, which are very helpful for improving the solubility of CNTs in water. Chitosan, a hydrophilic biopolymer, possesses three kinds of functional groups *viz.* amino, primary, and secondary hydroxyl groups in a glucosamine unit, and the functionalized CNTs contain carboxylic and hydroxyl groups.

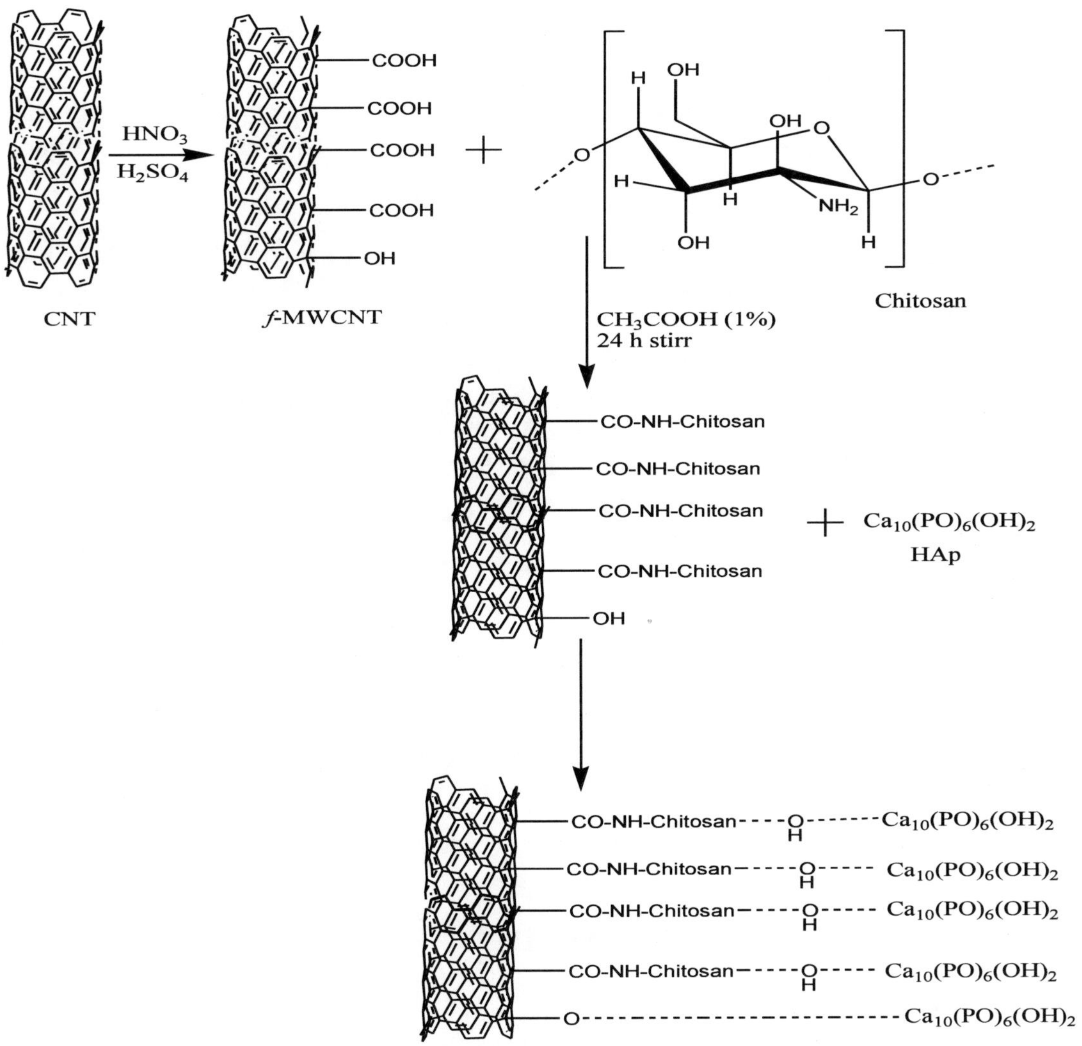

Figure 10. Chemical interactions between chitosan, HAp and CNT.

There is great possibility that strong hydrogen bonds may form between chitosan and the CNTs (Figure 10). Moreover, the addition of HAp in the carbon nanotube and chitosan solution, there is a strong co-ordination chemical bone might be possible with Ca^{2+} of HAp. Thus, tricomponent materials could be rigid and expect that to mimic the mechanical properties.

The compatibility and strong interaction between CNTs fillers and the matrix greatly enhances the dispersion as well as the interfacial adhesion, thus significantly increasing the mechanical properties of the matrix [67]. The covalent functionalisation of chitosan and CNT improves the interaction between chitosan with CNT, reduces the damage of CNT and increases its mechanical properties.

9. Challenges and Future Directions

The organic portion plays a major role in bone implants, therefore their matrices are often hybrid in composition. Apart from chitosan, HAp and CNT, large numbers of synthetic polymers have been used for the preparation of composite materials for artificial bone. It was observed that they showed better results in the case of cell viability, cell proliferation, alkaline phosphate activity and mineralization assays [81]. Several synthetic polymers have been used to prepare artificial scaffolding materials but have a number of disadvantages. One of the major drawbacks of synthetic polymers is their degradation times which are half than that of a natural polymer. Nowadays, synthetic polymers like poly lactic acid [82], polyethylene glycol [83], polyhydroxyl ethyl methacrylate with carbon nano tubes [84], poly caprolactone [85] and polylactic-co-glycolic acid [86] are widely used for scaffold materials. The synthetic polymer not only increases the scaffold strength, but also acts as a drug delivery device to induce bioactivity. Whether synthetic polymers would be able to reinforce the mechanical properties of HAp is still a burning question.

Even though CNT have tremendous mechanical strength and are able to mimic the natural bone function, the uniform dispersion of CNT in the chitosan and HAp matrix still remains a challenge. CNTs are not easily dispersed and bundles form in the polymer matrix. CNT should also be in a very pure form for biomedical applications as metal-containing CNTs are toxic to cells, and the production of pure forms of CNTs is another challenge. The cytotoxity of CNT is still a controversial topic due to errors in cytotoxic analysis techniques, as reported earlier, and the toxicity of CNT might vary depending on the cell type, purity of the CNTs and functionalisation.

Conclusion

Composite chitosan based materials have been found to have a predominant role in bone tissue engineering in recent years. The combination of biocompatible polymers and bioresorbable ceramic materials can mimic the natural function of bone. Chitosan, HAp with carbon nanotube composites cloud be potential bone implant materials with good osteo-conductive, osteoinductive and osteogenic properties. The structural, mechanical, chemical interaction and *in vitro* study of chitosan, HAp and CNT have been carried out for bone tissue engineering. Although many chitosan/HAp composite materials have been developed, problems persist with their mechanical properties. Hence, much research is in progress to address the gap in the development of mechanical properties. Much research work is still needed to address the cytotoxicity of CNTs. Though challenges still exist, the addition of CNT to improve the mechanical properties of chitosan and hydroxyapatite composite would surely support and stimulate the function of natural bone. The development of research on the efficacy of chitosan composite will open great possibilities for future bone tissue engineering.

REFERENCES

[1] Pizzoferrato, A; Cenni, E; Ciapetti, G; Granchi, D; Savarino, L; Stea, S. Inflammatory response to metals and ceramics. *Integrated Biomaterials Science. New York: Kluwer Academic/Plenum Publishers,* 2002 735–791.

[2] Wagh, A (2004): Chemically bonded phosphate ceramics: twenty-first century materials with diverse applications, Elsevier Science.

[3] Giannoudis, P; Dinopoulos, H; Tsiridis, E. Bone substitutes: an update. *Injury,* 2005 36, 20-27.

[4] Wang, M. Developing bioactive composite materials for tissue replacement. *Biomaterials,* 2003 24, 2133-2151.

[5] Venkatesan, J; Kim, S-K. Chitosan composites for bone tissue engineering—An overview. *Marine Drugs,* 2010 8, 2252-2266.

[6] Venkatesan, J; Kim, S K. Effect of temperature on isolation and characterization of hydroxyapatite from tuna (*Thunnus obesus*) bone. *Materials,* 2010 3, 4761-4772.

[7] Venkatesan, J; Qian, Z-J; Ryu, B; Ashok Kumar, N; Kim, S-K. Preparation and characterization of carbon nanotube-grafted-chitosan - Natural hydroxyapatite composite for bone tissue engineering. *Carbohydrate Polymers,* 2011 83, 569-577.

[8] Venkatesan, J; Qian, Z-J; Ryu, B; Noel, V T; Kim, S-K. A comparative study of thermal calcination and an alkaline hydrolysis method in the isolation of hydroxyapatite from Thunnus obesus bone. *Biomedical Materials,* 2011 6, 035003.

[9] Weiner, S; Wagner, H D. The material bone: structure-mechanical function relations. *Annual Review of Materials Science,* 1998 28, 271-298.

[10] Gay, C; Gilman, V; Sugiyama, T. Perspectives on osteoblast and osteoclast function. *Poultry Science,* 2000 79, 1005.

[11] Weiner, S; Wagner, H D. The materials bone: Structure-mechanical function relations. *Annual Review of Materials Science,* 1998 28, 271-298.

[12] Langer, R; Vacanti, J. Tissue engineering. *Science,* 1993 260, 920-926.

[13] Ratner, B; Bryant, S. Biomaterials: where we have been and where we are going. *Biomedical Engineering,* 2004 6,

[14] Langer, R. Biomaterials in drug delivery and tissue engineering: one laboratory's experience. *Acc. Chem. Res,* 2000 33, 94-101.

[15] Jiang, D. R and d of advanced ceramics activities in china and shanghai institute of ceramics chinese academy of sciences (siccas). *Mechanical Properties and Performance of Engineering Ceramics and Composites IV,* 2009 3.

[16] Liu, X; Ma, P X. Polymeric scaffolds for bone tissue engineering. *Annals of Biomedical Engineering,* 2004 32, 477-486.

[17] Yoshikawa, H; Myoui, A. Bone tissue engineering with porous hydroxyapatite ceramics. *Journal of Artificial Organs,* 2005 8, 131-136.

[18] Di Martino, A; Sittinger, M; Risbud, M. Chitosan: a versatile biopolymer for orthopaedic tissue-engineering. *Biomaterials,* 2005 26, 5983-5990.

[19] Best, S M; Porter, a E; Thian, E S; Huang, J. Bioceramics: Past, present and for the future. *Journal of the European Ceramic Society,* 2008 28, 1319-1327.

[20] Rezwan, K; Chen, Q Z; Blaker, J J; Boccaccini, a R. Biodegradable and bioactive porous polymer/inorganic composite scaffolds for bone tissue engineering. *Biomaterials,* 2006 27, 3413-3431.

[21] Legeros, R Z. Calcium phosphate-based osteoinductive materials. *Chemical Reviews,* 2008 108, 4742-4753.

[22] Sopyan, I; Mel, M; Ramesh, S; Khalid, K. Porous hydroxyapatite for artificial bone applications. *Science and Technology of Advanced Materials,* 2007 8, 116-123.

[23] Palmer, L C; Newcomb, C J; Kaltz, S R; Spoerke, E D; Stupp, S I. Biomimetic systems for hydroxyapatite mineralization inspired by bone and enamel. *Chemical Reviews,* 2008 108, 4754-4783.

[24] Koutsopoulos, S. Synthesis and characterization of hydroxyapatite crystals: A review study on the analytical methods. *Journal of biomedical materials research,* 2002 62, 600-612.

[25] Tan, Q; Zhang, K; Gu, S; Ren, J. Mineralization of surfactant functionalized multi-walled carbon nanotubes (MWNTs) to prepare hydroxyapatite/MWNTs nanohybrid. *Applied Surface Science,* 2009 255, 7036-7039.

[26] Jarudilokkul, S; Tanthapanichakoon, W; Boonamnuayvittaya, V. Synthesis of hydroxyapatite nanoparticles using an emulsion liquid membrane system. *Colloids and Surfaces A: Physicochemical and Engineering Aspects,* 2007 296, 149-153.

[27] Sarig, S; Kahana, F. Rapid formation of nanocrystalline apatite. *Journal of Crystal Growth,* 2002 237-239, 55-59.

[28] Xu, J L; Khor, K A; Dong, Z L; Gu, Y W; Kumar, R; Cheang, P. Preparation and characterization of nano-sized hydroxyapatite powders produced in a radio frequency (rf) thermal plasma. *Materials Science and Engineering A,* 2004 374, 101-108.

[29] Cao, L-Y; Zhang, C-B; Huang, J-F. Synthesis of hydroxyapatite nanoparticles in ultrasonic precipitation. *Ceramics International,* 2005 31, 1041-1044.

[30] Guo, G; Sun, Y; Wang, Z; Guo, H. Preparation of hydroxyapatite nanoparticles by reverse microemulsion. *Ceramics International,* 2005 31, 869-872.

[31] Feng, W; Mu-Sen, L; Yu-Peng, L; Yong-Xin, Q. A simple sol-gel technique for preparing hydroxyapatite nanopowders. *Materials Letters,* 2005 59, 916-919.

[32] Tseng, Y-H; Kuo, C-S; Li, Y-Y; Huang, C-P. Polymer-assisted synthesis of hydroxy-apatite nanoparticle. *Materials Science and Engineering: C,* 2009 29, 819-822.

[33] Pallela, R; Venkatesan, J; Kim, S K. Polymer assisted isolation of hydroxyapatite from Thunnus obesus bone. *Ceramics International,* 2011 37 3489-3497.

[34] Chai, C; Leong, K W. Biomaterials approach to expand and direct differentiation of stem cells. *Molecular Therapy,* 2007 15, 467-480.

[35] Seong, J M; Et Al. Stem cells in bone tissue engineering. *Biomedical Materials,* 2010 5, 062001.

[36] Ehrlich, H. Chitin and collagen as universal and alternative templates in biomineralization. *International Geology Review,* 2010 52, 661 - 699.

[37] Kim, S; Mendis, E. Bioactive compounds from marine processing byproducts-A review. *Food research international,* 2006 39, 383-393.

[38] Kim, S; Rajapakse, N. Enzymatic production and biological activities of chitosan oligosaccharides (COS): A review. *Carbohydrate Polymers,* 2005 62, 357-368.

[39] Kim, S (2010): Chitin, chitosan, oligosaccharides and their derivatives: biological activities and applications, Taylor and Francis.

[40] Liu, X; Lim, S. Templating and supersaturation-driven anti-templating: principles of biomineral architecture. *J. Am. Chem. Soc,* 2003 125, 888-895.

[41] Landis, W; Silver, F; Freeman, J. Collagen as a scaffold for biomimetic mineralization of vertebrate tissues. *Journal of Materials Chemistry,* 2006 16, 1495-1503.

[42] Lee, C; Singla, A; Lee, Y. Biomedical applications of collagen. *International journal of pharmaceutics,* 2001 221, 1-22.

[43] Wang, W; Itoh, S; Aizawa, T; Okawa, A; Sakai, K; Ohkuma, T; Demura, M. Development of an injectable chitosan/marine collagen composite gel. *Biomedical Materials,* 2010 5, 065009.

[44] Pallela, R; Venkatesan, J; Janapala, V R; Kim, S-K. Biophysicochemical evaluation of chitosan-hydroxyapatite-marine sponge collagen composite for bone tissue engineering. *Journal of Biomedical Materials Research Part A,* 2011 n/a-n/a.

[45] Ramshaw, J; Peng, Y; Glattauer, V; Werkmeister, J. Collagens as biomaterials. *Journal of Materials Science: Materials in Medicine,* 2009 20, 3-8.

[46] Di Martino, A; Sittinger, M; Risbud, M V. Chitosan: A versatile biopolymer for orthopaedic tissue-engineering. *Biomaterials,* 2005 26, 5983-5990.

[47] Teng, S; Lee, E; Yoon, B; Shin, D; Kim, H; Oh, J. Chitosan/nanohydroxyapatite composite membranes via dynamic filtration for guided bone regeneration. *Journal of Biomedical Materials Research Part A,* 2009 88, 569-580.

[48] Hu, Q; Li, B; Wang, M; Shen, J. Preparation and characterization of biodegradable chitosan/hydroxyapatite nanocomposite rods via in situ hybridization: a potential material as internal fixation of bone fracture. *Biomaterials,* 2004 25, 779-785.

[49] Chen, F; Wang, Z; Lin, C. Preparation and characterization of nano-sized hydroxyapatite particles and hydroxyapatite/chitosan nano-composite for use in biomedical materials. *Materials Letters,* 2002 57, 858-861.

[50] Zhang, Y; Venugopal, J R; El-Turki, A; Ramakrishna, S; Su, B; Lim, C T. Electrospun biomimetic nanocomposite nanofibers of hydroxyapatite/chitosan for bone tissue engineering. *Biomaterials,* 2008 29, 4314-4322.

[51] Xianmiao, C; Yubao, L; Yi, Z; Li, Z; Jidong, L; Huanan, W. Properties and in vitro biological evaluation of nano-hydroxyapatite/chitosan membranes for bone guided regeneration. *Materials Science and Engineering: C,* 2009 29, 29-35.

[52] Kong, L; Gao, Y; Cao, W; Gong, Y; Zhao, N; Zhang, X. Preparation and characterization of nano-hydroxyapatite/chitosan composite scaffolds. *Journal of Biomedical Materials Research Part A,* 2005 75A, 275-282.

[53] Manjubala, I; Ponomarev, I; Wilke, I; Jandt, K. Growth of osteoblast-like cells on biomimetic apatite-coated chitosan scaffolds. *Journal of Biomedical Materials Research Part B: Applied Biomaterials,* 2008 84, 7-16.

[54] Thein-Han, W; Misra, R. Biomimetic chitosan–nanohydroxyapatite composite scaffolds for bone tissue engineering. *Acta Biomaterialia,* 2009 5, 1182-1197.

[55] Oliveira, J; Rodrigues, M; Silva, S; Malafaya, P; Gomes, M; Viegas, C; Dias, I; Azevedo, J; Mano, J; Reis, R. Novel hydroxyapatite/chitosan bilayered scaffold for osteochondral tissue-engineering applications: Scaffold design and its performance when seeded with goat bone marrow stromal cells. *Biomaterials,* 2006 27, 6123-6137.

[56] Ding, S. Biodegradation behavior of chitosan/calcium phosphate composites. *Journal of Non-Crystalline Solids,* 2007 353, 2367-2373.

[57] Yamaguchi, I; Tokuchi, K; Fukuzaki, H; Koyama, Y; Takakuda, K; Monma, H; Tanaka, J. Preparation and microstructure analysis of chitosan/hydroxyapatite nanocomposites. *Journal of biomedical materials research,* 2001 55, 20-27.

[58] Kim, S-K; Mendis, E. Bioactive compounds from marine processing byproducts - A review. *Food Research International,* 2006 39, 383-393.

[59] Nath, S; Dey, A; Mukhopadhyay, a K; Basu, B. Nanoindentation response of novel hydroxyapatite-mullite composites. *Materials Science and Engineering: A,* 2009 513-514, 197-201.

[60] Saito, N; Aoki, K; Usui, Y; Shimizu, M; Hara, K; Narita, N; Ogihara, N; Nakamura, K; Ishigaki, N; Kato, H; Haniu, H; Taruta, S; Ahm Kim, Y; Endo, M. Application of carbon fibers to biomaterials: A new era of nano-level control of carbon fibers after 30-years of development. *Chemical Society Reviews,* 2011

[61] Iijima, S. Helical microtubules of graphitic carbon. *Nature,* 1991 354, 56-58.

[62] Tan, H; Chu, C; Payne, K; Marra, K. Injectable in situ forming biodegradable chitosan-hyaluronic acid based hydrogels for cartilage tissue engineering. *Biomaterials,* 2009 30, 2499-2506.

[63] Nair, L; Laurencin, C. Biodegradable polymers as biomaterials. *Progress in polymer science,* 2007 32, 762-798.

[64] Arpornmaeklong, P; Pripatnanont, P; Suwatwirote, N. Properties of chitosan-collagen sponges and osteogenic differentiation of rat-bone-marrow stromal cells. *International journal of oral and maxillofacial surgery,* 2008 37, 357-366.

[65] Ajayan, P; Zhou, O. Applications of carbon nanotubes. *Carbon nanotubes,* 2001 391-425.

[66] Samal, S; Bal, S. Carbon nanotube reinforced ceramic matrix composites-A review. *Journal of Minerals and Materials Characterization and Engineering,* 2008 7, 355-370.

[67] Wang, S-F; Shen, L; Zhang, W-D; Tong, Y-J. Preparation and mechanical properties of chitosan/carbon nanotubes composites. *Biomacromolecules,* 2005 6, 3067-3072.

[68] Usui, Y; Aoki, K; Narita, N; Murakami, N; Nakamura, I; Nakamura, K; Ishigaki, N; Yamazaki, H; Horiuchi, H; Kato, H. Carbon nanotubes with high bone-tissue compatibility and bone-formation acceleration effects. *small,* 2008 4, 240.

[69] Terada, M; Abe, S; Akasaka, T; Uo, M; Kitagawa, Y; Watari, F. Development of a multiwalled carbon nanotube coated collagen dish. *Dental Materials Journal,* 2009 28, 82-88.

[70] Liu, Y; Tang, J; Chen, X; Xin, J H. Decoration of carbon nanotubes with chitosan. *Carbon,* 2005 43, 3178-3180.

[71] Lau, C; Cooney, M J; Atanassov, P. Conductive macroporous composite chitosan –carbon nanotube scaffolds. *Langmuir,* 2008 24, 7004-7010.

[72] Qiu, J-D; Xie, H-Y; Liang, R-P. Preparation of porous chitosan/carbon nanotubes film modified electrode for biosensor application. *Microchimica Acta,* 2008 162, 57-64.

[73] Kumar, B; Feller, J-F; Castro, M; Lu, J. Conductive bio-polymer nano-composites (CPC): Chitosan-carbon nanotube transducers assembled via spray layer-by-layer for volatile organic compound sensing. *Talanta,* 2010 81, 908-915.

[74] Ozarkar, S; Jassal, M; Agrawal, A. pH and electrical actuation of single walled carbon nanotube/chitosan composite fibers. *Smart Materials and Structures,* 2008 17, 055016.

[75] Spinks, G M; Shin, S R; Wallace, G G; Whitten, P G; Kim, S I; Kim, S J. Mechanical properties of chitosan/CNT microfibers obtained with improved dispersion. *Sensors and Actuators B: Chemical,* 2006 115, 678-684.

[76] Shokrgozar, M A; Mottaghitalab, F; Mottaghitalab, V; Farokhi, M. Fabrication of porous chitosan/poly (vinyl alcohol) reinforced single-walled carbon nanotube nanocomposites for neural tissue engineering. *Journal of Biomedical Nanotechnology,* 2011 7, 276-284.

[77] Liao, H; Qi, R; Shen, M; Cao, X; Guo, R; Zhang, Y; Shi, X. Improved cellular response on multiwalled carbon nanotube-incorporated electrospun polyvinyl alcohol/chitosan nanofibrous scaffolds. *Colloids and Surfaces B: Biointerfaces,* 2011 84, 528-535.

[78] Zanello, L P; Zhao, B; Hu, H; Haddon, R C. Bone cell proliferation on carbon nanotubes. *Nano Letters,* 2006 6, 562-567.

[79] Zhang, X; Meng, L; Lu, Q. Cell behaviors on polysaccharide-wrapped single-wall carbon nanotubes: A auantitative study of the surface properties of biomimetic nanofibrous scaffolds. *ACS Nano,* 2009 3, 3200-3206.

[80] Wörle-Knirsch, J M; Pulskamp, K; Krug, H F. Oops They Did It Again! Carbon Nanotubes Hoax Scientists in Viability Assays. *Nano Letters,* 2006 6, 1261-1268.

[81] Spitalsky, Z; Tasis, D; Papagelis, K; Galiotis, C. Carbon nanotube-polymer composites: Chemistry, processing, mechanical and electrical properties. *Progress in Polymer Science,* 2010 35, 357-401.

[82] Yanoso-Scholl, L; Jacobson, J A; Bradica, G; Lerner, a L; O'keefe, R J; Schwarz, E M; Zuscik, M J; Awad, H A. Evaluation of dense polylactic acid/beta-tricalcium phosphate scaffolds for bone tissue engineering. *Journal of Biomedical Materials Research Part A,* 2010.

[83] Ma, G; Yang, D; Li, Q; Wang, K; Chen, B; Kennedy, J F; Nie, J. Injectable hydrogels based on chitosan derivative/polyethylene glycol dimethacrylate/N, N-dimethylacrylamide as bone tissue engineering matrix. *Carbohydrate Polymers,* 2010 79, 620-627.

[84] Ashok Kumar, N; Ganapathy, H S; Kim, J S; Jeong, Y S; Jeong, Y T. Preparation of poly 2-hydroxyethyl methacrylate functionalized carbon nanotubes as novel biomaterial nanocomposites. *European Polymer Journal,* 2008 44, 579-586.

[85] Binulal, N; Deepthy, M; Selvamurugan, N; Shalumon, K; Suja, S; Mony, U; Jayakumar, R; Nair, S. Role of nanofibrous poly (caprolactone) scaffolds in human mesenchymal stem cell attachment and spreading for in vitro bone tissue engineering—response to osteogenic regulators. *Tissue Engineering Part A,* 2010 16, 393-404.

[86] Hafezi-Ardakani, M; Kavian, F; Moztarzadeh, F; Eslaminejad, M B; Zamanian, A; Bagheri, F. Poly (lactic-co-glycolic)/nanostructured merwinite porous composites for bone tissue engineering. I. Preparation and morphology. *Key Engineering Materials,* 2012 493, 718-722.

In: Tissue Engineering
Editors: T. Ishikawa and M. Yamamoto
ISBN 978-1-62257-062-1

Chapter 3

STRETCHING BIOREACTOR FOR DYNAMIC ENGINEERING OF MUSCLE TISSUES

Marie-Noëlle Giraud[1], Dominic Bertschi[2], Anne Géraldine Guex[2, 4], Céline Fouassier[2], Fabrizio Romano[2], Silvio Näther[3], Giuseppino Fortunato[4], Thierry P. Carrel[2] and Hendrik T. Tevaearai[2]

[1]Cardiology, Department of Medicine, University of Fribourg, Fribourg, Switzerland
[2]Cardiovascular Surgery, Inselspital and University of Berne, Berne, Switzerland
[3]Center of Forensic Imaging and Virtopsy, University of Berne, Berne, Switzerland
[4]EMPA, Swiss Federal Laboratories for Materials Science and Technology, St Gallen, Switzerland

ABSTRACT

Objectives: Mechanical stimulation of engineered tissue holds great promise for the development of functionally structured tissues. In particular, mimicking the mechanical properties of native heart and the loading conditions of the cardiac tissue in vitro relies on dynamic, three dimensional culture systems that simulate the anatomical geometry and phasic contraction of cardiac left ventricles (LV). We report our study on the development and validation of a new device composed of a LV shaped muscle biograft built on electrospun fibers submitted to circumferential stretch.

Methods: Custom made silicone left ventricle (LV) shaped bulbs were produced with a water-soluble wax mold and covered with an electrospun poly(ε-caprolactone) (PCL) micron-scaled fibrous matrix. Pump controlled volumetric changes induced bulb enlargement and resulted in matrix stretching. Spatial characterization of the stretch was analyzed using a GOM 3D digitizer and GOM ARAMIS software. C_2C_{12} cells were seeded on the matrix and cultured for 1 week under mechanical stimulation. Cyclic (1Hz) and ramp (cycles of 24h stretch /24h rest) stretches were applied. Cellular distribution was investigated by DNA quantification and tissue thickness evaluated using 3D confocal microscopy.

Results: A gradient of surface strains was characterized from the base to the apex of the bulb. When inflated, the apex showed a linear increase in the strain from 4 to 24%. Meanwhile, the increase of strain at the base was limited and ranged from 0% to 5%. Dynamic culture induced variations in cell density and tissue thickness. This effect appeared to be dependent on the applied stretch amplitude.

Conclusion: We provide evidence that our new device composed of a LV shaped bulb covered with a microfibrous matrix is promising for structured muscle tissue generation and will further generate a broad interest from researchers investigating the effect of anisotropic and circumferential stretch on cardiac stem cells and cardiac physiopathology.

1. Introduction

The use of electrospun fibers for tissue engineering has gained increasing interest. Electrospinning allows the engineering of tailored fibers of various materials with controlled architecture. Using these characteristics, many efforts have been undertaken to mimic a natural extracellular matrix (ECM) in terms of structure, topography and composition. A wide range of polymers, fiber sizes and orientation as well as complex 3D environments have been engineered [1-5]. Subsequently, pluridisciplinary approaches enabled the investigation of the cellular organization and differentiation in relation to physical and chemical properties of selected biomaterials [6]. These numerous studies gave important insight into mechanisms underlying biomaterials-cell interaction. However, these studies often lacked the implementation of external stimuli such as mechanical input. It is well established that mechanical stresses such as compression, tension or shear regulate ECM synthesis [7] and trigger cellular organization and differentiation [8]. Accordingly, the amelioration of the structure and function of engineered tissue may greatly depend on the implementation of mechanical stimulation during tissue culture. In addition, mimicking the shape of the target tissue is of particular importance as it allows the investigation of the tissue formation in response to local mechanical signals [9; 10]. For instance, it has been shown to be relevant for tissue engineering of cardiac valves using pre-shaped scaffold mimicking valve leaflets [11]. In particular, Lewinsohn et al [12] combined stretch with specific scaffold shape. The authors highlighted the importance of the tissue structure and its implication of local mechanical environment and its subsequent influence of tissue function. They provided evidence that circumferential loading was most favorable for cardiac valve function. They demonstrated that local strain varied with axial or circumferential type of strain with a major effect on matrix reorganization. So far, the prospect of designing scaffolds according to the anatomical geometry of left ventricle has not yet been investigated.

Cells have the ability to actively respond to their mechanical microenvironment in term of adhesion, proliferation, morphology, ECM secretion and gene expression profile. In order to elucidate the effects of different mechanical stimuli, different type and combination of dynamic stretch conditions may be applied. Rapid cyclic stress that reflects cardiac beats has been shown to induce pronounced effects on most cell behavior. A slower stretch pattern such as ramp stretch with series of static stretch periods followed by resting phases may have a different stimulation effect. For cells such as mesenchymal stem cells and fibroblasts, static stretch usually induced less pronounced alterations than dynamic stretch. On the contrary, both stretch types induced hypertrophy of cardiomyocytes [10]. Ramp stretch represents slow dynamic loading conditions that may differentially alter cell response.

Accordingly, we focus on the development of a unique system composed of a bulb mimicking the dynamic three-dimensional aspects of the LV as a carrier for the engineered tissue based on a microfibrous scaffold. The engineered tissue was submitted to dynamic culture conditions specifically cyclic or ramp stretches. We validated this new dynamic cell culture system using a myoblast cell line.

2. Materials and Methods

2.1. LV Shaped Bulb Production

We developed a LV shaped bulb covered with electrospun fibers. The manufacture involved a three-step procedure (Figure 1). Following the original work of Gonen-Wadmany et al [13] we used an inflatable silicone bulb for the creation of circumferential load. Our polymeric fibrous matrix was then deposited at the surface of the silicone bulb using an electrospinning process as briefly described below.

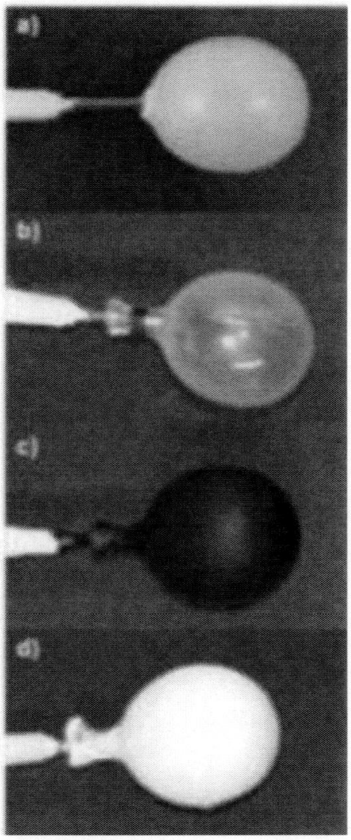

Figure 1. LV shaped bulb production. LV shaped culture substrates were produced in a stepwise method. A negative wax mold (a) was first coated by a two-component silicone (b), followed by a carbon black enriched silicone layer (c). Finally, silicone bulbs were coated with an electrospun matrix of PCL (d).

2.1.1. Preparation of Negative Wax-Molds

In a first step, a water-soluble wax mold (Sol-U-Carve wax, Freeman, USA) served as a template. Water-soluble wax was melted at 90°C in a water bath and then injected into a rubber master mold. The wax was left to set and dry at 4°C for 15 minutes.

2.1.2. Silicone Coating

Secondly, 6 wax molds were fixed on a rack, dipped in liquid two-component silicone (ALPA-Sil 32, Silitech AG Worb, Switzerland) and dried overnight. Next, the wax core was dissolved to obtain an inflatable silicone bulb. The bulbs were then fixed on a 15-gauge luer stub adapter, connected to a luer-lock valve system (QosinaCoorp. USA). Silicone bulbs were then coated with a thin layer of carbon black enriched two-component silicone (Silitech AG Worb, Switzerland) and dried at room temperature for 48 hours. Balloons were rinsed in ethanol and filled with 4 mL sterile water.

2.1.3. Matrix Deposition by Electrospinning

Poly(ε-caprolactone) (PCL, 70'000 to 90'000 $gmol^{-1}$), chloroform (CHCl3, ≥99%) and methanol (CH3OH, ≥99%) were obtained from Sigma-Aldrich, Switzerland. 15 % w/v PCL spinning solutions were prepared by dissolving PCL in a mixture of chloroform/methanol (9:1). Briefly, the electrospinning apparatus consisted of an infusion pump (KD scientific, USA) providing a constant polymer flow through a plastic syringe (5 mL, BD, USA) equipped with a stainless steel needle of an inner diameter of 0.8 mm (Unimed S.A., Switzerland). A positive (needle) and a negative (bulb) voltage supply source (aip Wild AG, Switzerland) provided the electrical field strength. Silicone bulbs were suspended on a tilted rotatory device. All equipments were placed in a grounded Faraday cage and a chemical hood. Fibers were spun at a flow rate of 50 $\mu Lmin^{-1}$, applying field strength of 1 $kVcm^{-1}$. The bulbs were homogenously coated by constantly rotating the bulb and adjusting the tilt angle of the fixation.

2.2. LV Shaped Bulb Characterization

2.2.1. Scanning Electron Microscopy

Fiber morphology was characterized by scanning electron microscopy (SEM). 1 cm^2 sections of the electrospun matrix were removed from the bulb and plasma gold sputtered (Polaron Equipment, SEM coating Unit E5100, Kontron AG, Switzerland; 5 mA, 1 mbar, 7 min) prior to imaging. Micrographs were recorded on a Hitachi S-4800 (Hitachi High-Technologies, Canada) at an accelerating voltage of 2 kV and 10 μA current flow. For fiber diameter measurements, 200 individual values were evaluated. Matrix thickness was measured on a horizontally suspended matrix section.

2.2.2. Strain Characterization

Characterization of motion distribution and regional strain at the surface of the LV shaped bulb was performed using a 3D digitizer camera and the software ARAMIS (GOM, GesellschaftfüroptischeMesstechink, Germany). The LV shaped bulb was sprayed with grey paint in order to create a grey pattern at the surface.

The bulbs were then dipped in a water bath and placed in front of the 3D digitizer camera. The procedure was carried with different filling volumes: the bulb was inflated by a stepwise addition of 0.5 mL water from 4.5 mL up to 8 mL. After each volume change, the 3D coordinates (x,y,z) for each camera pixel were recorded.

Based on this 3-dimensional imaging of the inflatable bulb, the motion distribution and regional strain were quantified using GOM ARAMIS software. A selection of 5 points situated at the apex, base and equator of the bulb were selected and the quantified surface strain at each point was graphically represented in function of filling volume. Each measurement was repeated on 3 different bulbs.

2.3. Cell Culture

2.3.1. Culture Chamber

Custom designed culture chamber (produced at Traboldand Co. AG Berne, Switzerland) allowed simultaneous culture of three bulbs in one cylindrical culture chamber. It consisted of two main parts (Figure 2): a 800 mL cylindrical glass-pot and the matching cover plate where the bulbs were fixed and connected to the pumps through teflon units.

2.3.2. Pumps and Mechanical Stimulation

A pump was connected to the silicone bulbs to control volumetric changes and subsequent matrix stretching.

Figure 2. The bioreactor. A custom made 800 mL culture chamber allowed for the simultaneous culture of three bulbs. Bulbs were suspended on Teflon lids, immersed into culture medium and connected to pumps via a tubing system.

To study the cell response to the stretching of the carrier bulb, two different mechanical stimulation modes were defined (Figure 3):

- Slow ramp stretch mode with a filling time of 20 minutes and a volume change of 1 mL (from 4.5 to 5.5 mL) followed by a static strain time of 24 hours followed by an analogous emptying and rest time without strain application.
- Cyclic stretch mode with a fast volume change from 4.5 mL to 5.5 mL every minute, resulting in a stretch frequency of 1 Hz.

In both cases, stretching mode was started 24h after the seeding procedure and sustained for 7 days.

2.3.3. Cell Expansion

Mouse skeletal myoblasts, cell line C_2C_{12} (CRL-1772, ATCC, USA) were cultured up to 80% confluency at 37°C in a humidified incubator with an atmosphere of 5% CO_2 and 95% air (Thermo Forma, Model 371, Thermo Fisher Scientific, USA). Cells were expanded in growth medium (HAM F-12 nutrient mixture medium, Gibco, Invitrogen, USA), 10% fetal bovine serum (FBS, PAA-Clone, PAA-laboratories, Austria), 1.25% penicillin/streptomycin solution (P/S, Gibco, Invitrogen, USA), 5 $ngmL^{-1}$ basic Fibroblast Growth Factor (bFGF, Sigma-Aldrich, Switzerland) and 40 μgmL^{-1}Dexamethasone (Dex, Sigma-Aldrich, Switzerland).

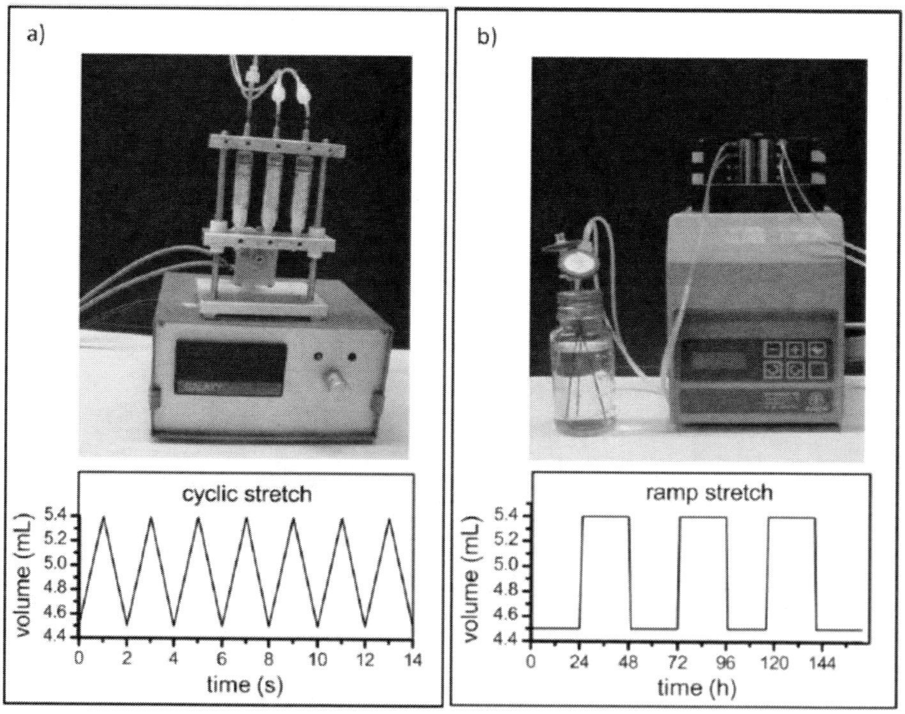

Figure 3. The pumps. Two distinct pumps allowed mechanical stimulation of the myoblasts with two different stretching patterns. A) cyclic stretch of 1 Hz, B) ramp stretch of a 24-hours frequency.

2.3.4. Cell Seeding

Cell seeding on the LV shaped bulb was performed in a cylindrical custom made seeding glass flask filled with growth medium and maintained at 37°C. An initial cell quantity of $5*10^6$ cells were added followed by 3 further injections of $5*10^6$ cells every 20 minutes. The flask was maintained horizontally and rotated 90° after each cell injection. A total cell quantity of $20*10^6$ cells was seeded per bulb.

The stepwise cell seeding was optimized and guaranteed a homogenous cell seeding on the whole bulb. The bulb was then removed carefully from the seeding flask and suspended in the culture chamber, attached to the lid and connected to the pump. A medium change was undertaken on the third day.

2.4. Cell Response Characterization

2.4.1. MTT Staining

Cell seeding on the LV shaped bulb was optimized in order to obtain a homogenous coverage of the entire bulb. Optimization steps were monitored by MTT staining. Seeded bulbs were incubated in growth medium, supplemented with 0.5 $mgmL^{-1}$Thiozolyl Blue Tetrazolium Bromide (MTT, Sigma-Aldrich, Switzerland) for 2h at 37°C. Subsequently, seeding efficiency was assessed qualitatively, based on observation of the distribution of blue MTT staining on the bulb.

2.4.2. Sampling

After 1 week in culture, bulbs were removed from the culture chamber and the PCL layer was carefully removed from the silicone bulbs with scissors. A slice was cut out of the scaffold and divided into 4 sections.

2.4.3. Cell Quantification Assay

Cell number was assessed by DNA quantification. Using a commercially available cell proliferation assay kit (CyQuant, Invitrogen, USA), total DNA amount was quantified by spectrofluorescence. Briefly, constructs were incubated in cell lysis buffer, sonicated (Sonopuls HD 2070/UW 2070, Bandelin electronic, Germany), vortexed and centrifuged. Supernatants were incubated with a DNA binding fluorescent agent (CyQuant GR dye solution). Quantification was accomplished by spectrophotometry (Inifite 200, Tecan Trading AG, Switzerland) at 480 nm excitation and 520 nm emission wavelengths.

The cell number was obtained using a standard curve. The area of each section was measured using Image J software (Wayne Rasband, United States National Institutes of Health, Bethesda, MD) and the density of cells calculated per section. The results per section are presented as a percentage of the entire slice.

2.4.4. Tissue Thickness

A fluorescent staining was performed to assess tissue thickness on the different positions of the bulb. Cells were fixed in 4% paraformaldehyde, prior to incubation in 0.2% Triton X-100 (Sigma-Aldrich, Switzerland).

Actin filaments were labeled with red Phalloidin–Tetramethylrhodamine B isothiocyanate (0.1 μM, 1:200 in PBS, Sigma-Alrich, Switzerland). Nuclei were counterstained with Hoechst bisBenzimide H33258 (1mgmL^{-1}, 1:200 in PBS, Sigma-Aldrich, Switzerland). Slides were mounted with Glycergel (Dako, USA). Tissue thickness was evaluated based on z-stack photographs, acquired on a confocal microscope (LSM 5 Exciter, Carl ZeissMicroImaging GmbH, Jena, Germany). Briefly, series of images were taken from the lowest part of the sample to the top, with a 0.3 μm intersectional distance.

Single images were assembled into 3D-reconstructional pictures, and thickness measurements were carried out at 16 different positions per section. Mean thickness was normalized to the maximal thickness for each experimental setting and the relative thickness expressed as a percentage of the control thickness for each section.

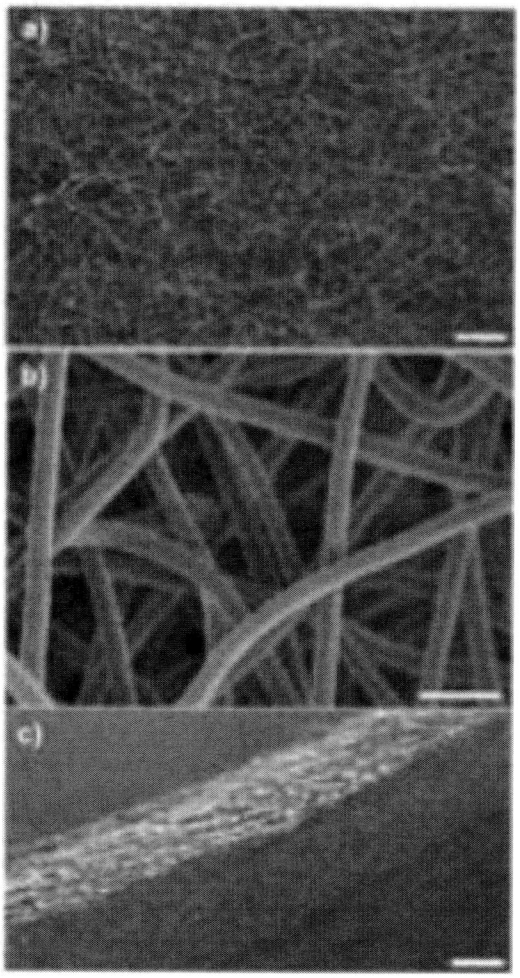

Figure 4. SEM images of electrospun fibers. Acquired SEM images of the deposited fibers illustrate the homogenous diameter distribution and coating of the matrix (a and b). Fibers displayed diameters of 2.3 ± 0.8 μm. On cross sectional images (c), the thickness was estimated to be 218 ± 50 μm. Scale bar 100, respectively 10 μm.

2.5. Statistical Analyses

All results are presented as mean value ± standard deviation; statistical analysis was performed using Statistica 9.1 (StatSoft, Inc., www.statsoft.com). An ANOVA-test was performed, followed by a Fisher LSD post hoc test. Differences were considered as significant for $p \leq 0.05$.

3. RESULTS

3.1. Electrospinning

The carbon black enriched silicone coating strongly increased the conductivity of the balloons and facilitated the fiber production. Measured voltage on the pure silicone balloon was 0 kV, whereas carbon black coated balloons were fully charged, represented by a measured voltage of 4 kV, as applied by the high voltage generator. Fiber deposition was stable and reproducible. SEM images confirmed the homogenous matrix (Figure 4), revealing fiber diameters of 2.3 ± 0.8 μm. Matrix thickness was assessed using cross sectional measurements. From base to apex a thickness of 218 ± 50 μm was measured.

3.2. Circumferential Strain Measurement

Surface strain of the bulb was quantified at different filling volumes when the LV shaped bulb was immerged in water. Figure 5 shows a color scale of the surface strain for a filling volume of 4.5 mL. The highest strains were located in the lower part of the bulb and varied from 10 to 20 %. Lower strains were recorded at the base of the bulb and varied from 0 up to 10%. Five different points were selected and the respective surface strains evaluated during the stepwise increase of the bulb volume from 0 mL to 7 mL as shown in Figure 5b. We observed a linear increase in the strain from 0 to 24% for the selected point 2, situated at the equator of the bulb. For point 1, situated at the base, the strain varied from 0 to a maximum of 5%. Interestingly, strain started to increase for a minimal filling volume of 1 mL.

3.3. Characterization of Cell Distribution

Seeding of the cells on a 3-dimensional structure is technically challenging. Using a rotational seeding device at 37°C, we obtained a homogenous cell distribution from base to apex of the bulb as shown by the uniform coloration of the LV shaped bulb after MTT staining of viable cells performed directly after the end of the seeding procedure (Figure 6). From apex to base, cells adhered at the surface of the matrix.

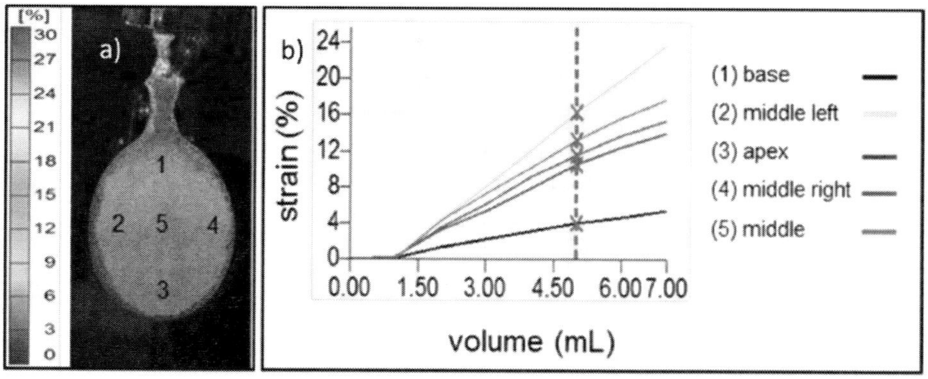

Figure 5. Strain patterns. 3-dimensional imaging of the inflatable bulb allowed quantification of the motion distribution and regional strain (GOM ARAMIS software). The measured strain, partitioned along the bulb is displayed as color-coded (a). The graph represents the surface strain at each selected point in function of filling volume (b). Lowest strain was recorded on the base of the bulb. Results were similar when assessed on 3 different bulbs.

C_2C_{12} were cultured for 7 days in the bioreactor under cyclic or ramp stretch together with a control bulb that underwent no stretch. C_2C_{12} spread at the surface of the PCL scaffold; no infiltration within the matrix was observed. After 7 days, the cell layer was not confluent for any of the conditions investigated. DNA quantification confirmed the homogenous distribution of the cell maintained over a 7-day period of culture.

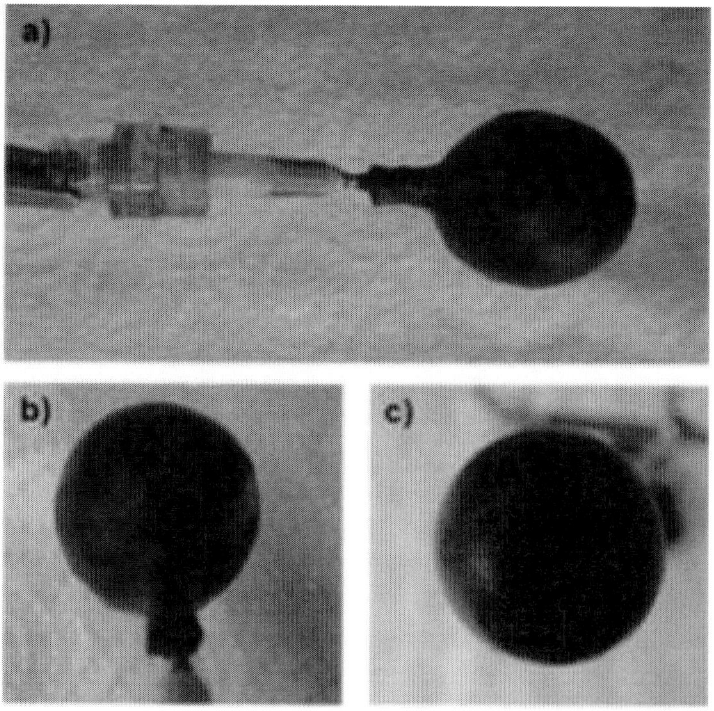

Figure 6. Seeding efficiency. MTT staining of viable cells allowed assessment of the cellular distribution among the bulb. As represented by the images, a homogenous seeding was achieved on the side (a), top (b) and bottom (c) of the bulb.

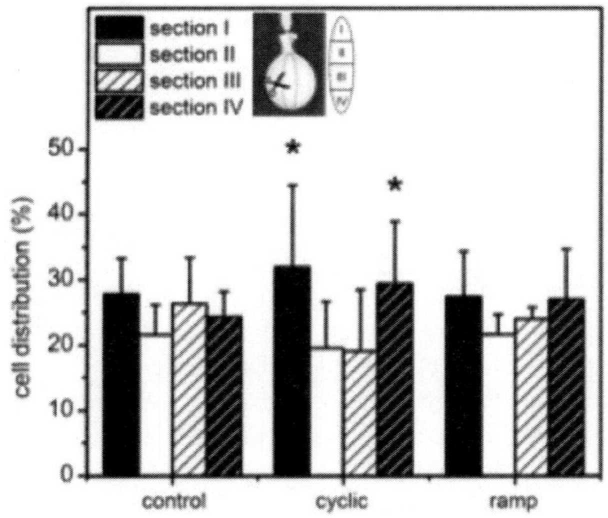

Figure 7. Cell quantification. Cell density was quantified based on DNA measurements related to surface area on different sections of the bulb. The density is expressed as relative to total cell number of the 4 sections. We showed that cells were equally distributed over each of the 4 sections in control and ramp stretch conditions. Cyclic stretch induced an alteration of the cell distribution. Numbers of cultured bulbs were 5, 4 and 6 respectively for unstretched (control), cyclic and ramp stretches. (*p≤0.05vs sections II and III).

The cells were equally distributed from section I to section IV for unstretched conditions (n=5) as shown in Figure 7. We also observed that this distribution stayed homogenous under ramp stretch conditions (n=6). By contrast, when cells were submitted to cyclic stretch, the middle parts (sections II and III) of the bulb presented significantly less cells compared to the apex and base (n=4).

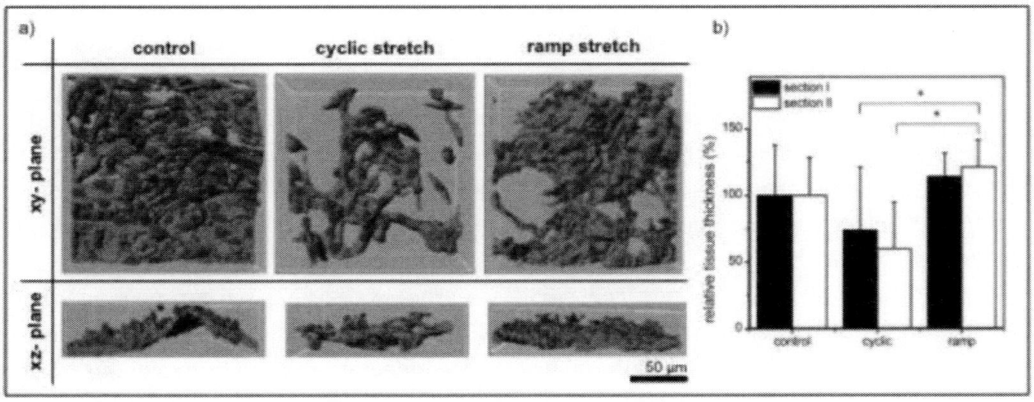

Figure 8. Estimated thickness. A) Tissue thickness was measured on fluorescently labeled constructs, based on 3D-reconstructional pictures (Imaris software). Red: actin skeleton, blue: nuclei. Thickness was measured at 16 different positions in each section. B). Thickness is expressed as a percentage relative to the unstretched condition (control). Highest thickness was found on constructs cultured under ramp stretching pattern or in control. Numbers of cultured bulbs were 5, 4 and 6 respectively for unstretched, cyclic and ramp stretches (*p≤0.05).

These results were confirmed by the measurements of tissue thickness. Figure 8 shows that the thickness of the tissue remained identical under ramp stretching compared to unstretched controls, whereas under cyclic stretch the thickness of the tissue was significantly reduced. A maximal thickness of 42±7 μm was obtained for a section II of a bulb's slide cultured under unstretched conditions.

4. Discussion

We developed a de novo bioreactor generating circumferential stretching of biocompatible matrix that mimics left ventricle anatomical geometry. We demonstrated that the surface strains could be estimated precisely for each point and varied from apex to the base of the bulb. In addition, our new device allowed the use of cyclic and ramp stretch conditions. We provided evidence of the feasibility of dynamic cell culture and showed that cell density could be altered by stretch patterns and strain amplitude. The design of two pumps that allow either cyclic fast stretch or ramp slow stretch with the possibility to vary the stretch pattern hold particular importance in the investigation of cells' response to different mechanical stimuli encountered during developmental, physiological or diseased conditions.

For this purpose, in vitro dynamic culture conditions that mimic mechanical properties of the native heart and loading conditions of the cardiac cell/tissue constitutes an important step toward modeling of cardiac pathophysiology as well as novel tools to engineered muscle tissue.

4.1. Electrospinning on Three Dimensional Devices

Electrospun fibers have various applications in skin engineering or wound dressings [1], engineered bone or cartilage [2; 3], muscle grafts [4], cardiac tissue [14] or vascular prostheses [15]. The large choice of available polymers allows for custom tailored chemical and mechanical properties, as well as distinct degradabilities. However, although being applied for various tissues of distinct shapes, electrospun substrates are mainly restricted to flat, macroscopically two-dimensional sheet-like structures. Various collectors, such as simple stainless steel plates [3] or drums rotating at different speed [2; 16] allowed the collection of randomly oriented or parallel oriented fibers respectively. So far, however, the deposition of electrospun fibers onto three-dimensional objects solely focused on the production of tubular scaffolds for vascular prostheses. Electrospun PCL fibers were collected on a rotating drum in order to produce small vascular grafts [17]. Direct coating of a 3 dimensional object, such as the LV shaped bulb, has however not been reported previously. The main challenge of coating a balloon lies in the generation of a homogenous matrix, covering the entire object. This is difficult to control as the electrospinning jet by its very nature describes a chaotic motion. This was mastered in our approach, by designing a conductive bulb, where a negative voltage could be applied, driving the polymer-jet towards the balloon. The balloon was then rotated and gradually tilted within the electrical field in order to guarantee uniform matrix deposition. In a first approach, we observed that the initiation of a rotating, non-conductive silicone balloon into an electrical field induced electrostatic instabilities, resulting in a deflected or disturbed polymer-jet. Fibers were directed away from the balloon and deposited heterogenously. We solved this challenge by introducing a conductive two-component silicone and illustrated that the conductive balloon allowed for reproducible and homogenous matrix coating and stable fiber formation. Furthermore, the high elasticity of the matrix enables the application of dynamic culture conditions on our construct.

4.2. Mechanical Stimulation

Myocardium sustains cyclic strain with magnitude and frequency related to heart rate and loading pressure. Nevertheless, myocardium undergoes different mechanical stimulations during development or pathophysiological conditions such as volume overload or myocardial infarction [18; 19]. Cardiac injury induces cell death followed by remodeling of the ventricle. Consequently, fibrosis, reduced wall thickness and changes in mechanical load modify mechanical stress applied to the myocardium. Furthermore, geometry modifications occur in pathological conditions, the ellipsoidal shape of the LV in normal conditions changes progressively to conical and spherical shapes [18].

These modifications, resulting in variations in mechanical strain, influence tissue remodeling and predominantly ECM synthesis [10]. Accordingly, the effect of different stretch patterns was investigated on major cell types found in the myocardium. It was demonstrated that each cell type responded differently to different stretch patterns. For instance, cardiomyocytes underwent hypertrophy when submitted to cyclic or static stretch [10] whereas fibroblasts increased both their secretion of collagen and release of growth factors [10]. These effects were more pronounced under cyclic stimulation than static tensile stress conditions [10]. Our results further showed the predominant effect of cyclic stretch on myoblast density compared to ramp stretch.

4.3. Circumferential Strain

The large majority of studies applied uni-or bi-axial tensile stress to 2D cultures. Conversely, 3D dynamic cultures with circumferential stretch is more likely to mimic native tissue but has rarely been investigated. The importance of circumferential stretch in heart pathophysiology has been defined by recent MRI left ventricle stain mapping. The circumvention strain varied from 25% shortening to 10% stretching in a normal left ventricle [20]. Circumferential strain maps of the left ventricle showed heterogeneous strain amplitude from base to apex but also from epical to endocardial layers [19; 21]. With a strain range between 0% and 24%, our inflatable device allows a wide range of possible strain application.

Circumferential stress was mainly applied to engineered vessels to mimic hypertension or to explore physiological exposition of the smooth muscle cells (SMC) to the pulsatile blood flow [22]. When SMC were embedded in hydrogel together with neonatal cardiomyocytes, circumferential stretch induced a strain magnitude dependent orientation of the cells and collagen fibrils [13]. Other recent studies underlining the importance of circumferential stretch explored its role for cardiac valves function and renewal [23]. The authors provided evidence that circumferential cyclic stretch increased the expression of a-SMA in porcine aortic valve leaflets. They therefore speculated that cyclic stretch is a regulator of the contractile and fibrotic phenotypes of aortic valve interstitial cells.

4.4. Importance of Heterogeneous Strain Simulation

A wide area of research on mechanotransduction relies on homogenous stretch. However, it is important to stress that the apparent homogeneity of the strains was often overestimated. Brown et al [24] provided evidence that unintentional heterogeneity in the strain was inherent to the flexible culture device commonly used in dynamic culture systems. The authors showed that substrate distension in these devices is often non-uniform and therefore creates an anisotropy in microenvironmental strains applied to the cells.

Moreover, in vivo, stretch distribution is often non-uniform. For instance, the myocardium has been defined as an active nonlinear anisotropic viscoelastic material [19]. In particular, circumferential strain in the left ventricle is heterogeneous and increasing sensibility and reliability in strain evaluation draw attention to micro-mechanical environment of the myocardium.

In addition, heterogeneity is increased after cardiac injury. Latimer et al [25] showed a complex distribution of stretch in the border zone surrounding the ischemic region and in non-ischemic regions surrounding the border zone.

Interestingly, the importance of anisotropic strain was initially established in plant cell biology. Using a gradient of compression, Lynch and Lintilhac [26] demonstrated the role of local strain in proplast orientation. Only recently, studies were undertaken to investigate the effect of the anisotropy of microenvironmental stretch influence eukaryotic cell behavior. For instance, this strain anisotropy has been shown to play an important role in cardiac valve function [12] and engineered cardiac valve generation [27]. These days, few new devices start to be designed to investigate the effect of cell gradient on cellular response as for example the flexible membrane with a rigid fixation at its center presented by Richardson et al [28]. Importantly, precise evaluation of mechanical anisotropy relies on precise strain measurement. We provide evidence that precise strain can be recorded and quantified for any selected point at the surface of the bulb using a 3D camera.

4.5. Cell Response

In the present study, we chose to validate our custom designed dynamic culture system using a myoblast cell line. We showed that cells adhere and spread on microfibers of the PCL matrix. The cells were homogenously distributed all over the LV shaped bulb after seeding as well as after one week of static culture. The major difficulty relies on the seeding setting and variability may occur if seeding parameters are not standardized.

Mechanical stimulation triggers skeletal muscle repair during exercise and the response of myoblasts has been explored extensively [29]. The satellite cells as muscle progenitor cells are activated by mechanical stress both in vivo and in vitro. Tatsumi et al [30] demonstrated that a brief period of stretch activated proliferation of quiescent primary satellite cells and presented the main trigger to a cascade of signaling resulting in the cell cycle initiation. In vitro, the response of myoblasts to stretch differs in function of the pattern of stretch and strain. Using uniaxial and multiaxial stretches, Hornberger at al [31] demonstrated that skeletal muscle cells have the fundamental ability to differentiate between different types of mechanical forces and identified differential signaling pathways. This concept of specificity in mechanotransduction for skeletal myoblasts could possibly explain the differences we observed between ramp and cyclic stretches.

Furthermore, we showed that cyclic stretch induced a reduction in cell density resulting in reduced tissue thickness at the site of increased strain. These results related to the investigation of Liu et al [32] that showed that magnitudes of cyclic-stretch from 6% induced apoptosis of cultured myoblasts.

4.6. Limitations

We developed a LV shaped matrix that mimics left ventricle geometry and allows the investigation of cell response after different strain stimulation. The present study validates the suitability of our 3D system to address dynamic interactions of circumferential stretch and cell matrix composition on the biology of seeded cells. However, further studies should be undertaken in order to elucidate specific cell response and correlate local strain with cell response.

Furthermore, cell confluency was not achieved after 1 week in culture. This may have an important impact on cell response to applied stretch. In order to increase cell density, a higher number of cells should be seeded or the culture time increased. Further investigation on the cell response according to a fine-tuned stretch pattern and cell density could give interesting insight.

Finally, a large quantity of cells was necessary to control the feasibility of our dynamic culture device. However, further studies that may involve cells with lower expansion abilities would benefit from the gradient of strains generated in our system. Indeed, low strain areas localized at the base of the LV shaped blub could possibly be considered as an internal reference for unstretched conditions.

Conclusion

Our new device enables the mechanical control of tissue formation and investigation of local cell response to circumferential strain. Further investigation will enable the definition of important key conditions in the generation of muscle biografts and improvement of cell density, cell organization and mechanical properties of the muscle construct. Cardiac stem cell differentiation under beating conditions is another potential field of investigation that may benefit from our device.

Acknowledgment

The authors would like to thank Claudia Haller and Alexander Panos for their collaboration as well as Laura Seidel for text editing and Prof. Martin Flueck for his advises during manuscript preparation. The project was supported by a grant from the Swiss Heart Foundation.

References

[1] Powell HM, Boyce ST (2009) Engineered human skin fabricated using electrospun collagen-PCL blends: morphogenesis and mechanical properties. *Tissue Eng. Part A* 15: 2177-2187.

[2] Thomas V, Jose MV, Chowdhury S, Sullivan JF, Dean DR, Vohra YK (2006) Mechano-morphological studies of aligned nanofibrous scaffolds of polycaprolactone fabricated by electrospinning. *J. Biomater. Sci. Polym.* Ed 17: 969-984.

[3] Li WJ, Tuli R, Okafor C, Derfoul A, Danielson KG, Hall DJ, Tuan RS (2005) A three-dimensional nanofibrous scaffold for cartilage tissue engineering using human mesenchymal stem cells. *Biomaterials* 26: 599-609.

[4] Riboldi SA, Sampaolesi M, Neuenschwander P, Cossu G, Mantero S (2005) Electrospun degradable polyesterurethane membranes: potential scaffolds for skeletal muscle tissue engineering. *Biomaterials* 26: 4606-4615.

[5] He W, Ma Z, Teo WE, Dong YX, Robless PA, Lim TC, Ramakrishna S (2009) Tubular nanofiber scaffolds for tissue engineered small-diameter vascular grafts. *J. Biomed. Mater. Res* A 90: 205-216.

[6] Nisbet DR, Forsythe JS, Shen W, Finkelstein DI, Horne MK (2009) Review paper: a review of the cellular response on electrospunnanofibers for tissue engineering. *J. Biomater. Appl.* 24: 7-29.

[7] Fluck M, Giraud MN, Tunc V, Chiquet M (2003) Tensile stress-dependent collagen XII and fibronectin production by fibroblasts requires separate pathways. *Biochim. Biophys. Acta* 1593: 239-248.

[8] Galbraith CG, Sheetz MP (1998) Forces on adhesive contacts affect cell function. *Curr. Opin. Cell Biol.* 10: 566-571.

[9] Berry JL, Steen JA, Koudy Williams J, Jordan JE, Atala A, Yoo JJ Bioreactors for development of tissue engineered heart valves. *Ann. Biomed. Eng.* 38: 3272-3279.

[10] Gupta V, Grande-Allen KJ (2006) Effects of static and cyclic loading in regulating extracellular matrix synthesis by cardiovascular cells. *Cardiovasc. Res.* 72: 375-383.

[11] Sacks MS, Schoen FJ, Mayer JE (2009) Bioengineering challenges for heart valve tissue engineering. *Annu. Rev. Biomed. Eng.* 11: 289-313.

[12] Lewinsohn AD, Anssari-Benham A, Lee DA, Taylor PM, Chester AH, Yacoub MH, Screen HR Anisotropic strain transfer through the aortic valve and its relevance to the cellular mechanical environment. *Proc. Inst. Mech. Eng.* H 225: 821-830.

[13] Gonen-Wadmany M, Gepstein L, Seliktar D (2004) Controlling the cellular organization of tissue-engineered cardiac constructs. *Ann. N. Y. Acad. Sci.* 1015: 299-311.

[14] Kai D, Prabhakaran MP, Jin G, Ramakrishna S Guided orientation of cardiomyocytes on electrospun aligned nanofibers for cardiac tissue engineering. *J. Biomed. Mater. Res. B. Appl. Biomater.* 98B: 379-386.

[15] Pektok E, Nottelet B, Tille JC, Gurny R, Kalangos A, Moeller M, Walpoth BH (2008) Degradation and healing characteristics of small-diameter poly(epsilon-caprolactone) vascular grafts in the rat systemic arterial circulation. *Circulation* 118: 2563-2570.

[16] Choi JS, Lee SJ, Christ GJ, Atala A, Yoo JJ (2008) The influence of electrospun aligned poly(epsilon-caprolactone)/collagen nanofiber meshes on the formation of self-aligned skeletal muscle myotubes. *Biomaterials* 29: 2899-2906.

[17] Nottelet B, Pektok E, Mandracchia D, Tille JC, Walpoth B, Gurny R, Moeller M (2009) Factorial design optimization and in vivo feasibility of poly(epsilon-caprolactone)-micro- and narofiber-based small diameter vascular grafts. *Journal of Biomedical Materials Research Part A* 89A: 865-875.

[18] Maffessanti F, Lang RM, Corsi C, Mor-Avi V, Caiani EG (2009) Feasibility of left ventricular shape analysis from transthoracic real-time 3-D echocardiographic images. *Ultrasound Med. Biol.* 35: 1953-1962.

[19] Sermesant M, Delingette H, Ayache N (2006) An electromechanical model of the heart for image analysis and simulation. *IEEE Trans. Med. Imaging* 25: 612-625.

[20] Osman NF, McVeigh ER, Prince JL (2000) Imaging heart motion using harmonic phase MRI. *IEEE Trans. Med. Imaging* 19: 186-202.

[21] Kim BS, Nikolovski J, Bonadio J, Mooney DJ (1999) Cyclic mechanical strain regulates the development of engineered smooth muscle tissue. *Nat. Biotechnol.* 17: 979-983.

[22] Stegemann JP, Nerem RM (2003) Phenotype modulation in vascular tissue engineering using biochemical and mechanical stimulation. *Ann. Biomed.* Eng. 31: 391-402.

[23] Balachandran K, Konduri S, Sucosky P, Jo H, Yoganathan AP (2006) An ex vivo study of the biological properties of porcine aortic valves in response to circumferential cyclic stretch. *Ann. Biomed. Eng.* 34: 1655-1665.

[24] Brown TD, Bottlang M, Pedersen DR, Banes AJ (1998) Loading paradigms--intentional and unintentional--for cell culture mechanostimulus. *Am. J. Med. Sci.* 316: 162-168.

[25] Latimer DC, Roth BJ, Parker KK (2003) Analytical model for predicting mechanotransduction effects in engineered cardiac tissue. *Tissue Eng.* 9: 283-289.

[26] Lynch TM, Lintilhac PM (1997) Mechanical signals in plant development: a new method for single cell studies. *Dev. Biol.* 181: 246-256.

[27] Courtney T, Sacks MS, Stankus J, Guan J, Wagner WR (2006) Design and analysis of tissue engineering scaffolds that mimic soft tissue mechanical anisotropy. *Biomaterials* 27: 3631-3638.

[28] Richardson WJ, Metz RP, Moreno MR, Wilson E, Moore JE A device to study the effects of stretch gradients on cell behavior. *J. Biomech. Eng.* 133: 101008.

[29] Fluck M (2006) Functional, structural and molecular plasticity of mammalian skeletal muscle in response to exercise stimuli. *J. Exp. Biol.* 209: 2239-2248.

[30] Tatsumi R, Liu X, Pulido A, Morales M, Sakata T, Dial S, Hattori A, Ikeuchi Y, Allen RE (2006) Satellite cell activation in stretched skeletal muscle and the role of nitric oxide and hepatocyte growth factor. *Am. J. Physiol. Cell Physiol.* 290: C1487-1494.

[31] Hornberger TA, Armstrong DD, Koh TJ, Burkholder TJ, Esser KA (2005) Intracellular signaling specificity in response to uniaxial vs. multiaxial stretch: implications for mechanotransduction. *Am. J. Physiol. Cell Physiol.* 288: C185-194.

[32] Liu J, Wang Y, Yuan X, Feng Y, Liu H Cyclic-stretch induces the apoptosis of myoblast by activation of Caspase-3 protease in a magnitude-dependent manner. *Int. J. Biochem. Cell Biol.* 42: 2004-2011.

In: Tissue Engineering
Editors: T. Ishikawa and M. Yamamoto
ISBN 978-1-62257-062-1

Chapter 4

STRATEGIES FOR INTEGRATION OF TISSUE ENGINEERED CONSTRUCTS TO NATIVE CARTILAGE AND BONE

Rupak Dua and Sharan Ramaswamy*
Tissue Engineered Mechanics,
Imaging and Materials (TEMIM) Laboratory,
Department of Biomedical Engineering,
Florida International University, Miami, Flordia, US

ABSTRACT

Tissue engineered constructs typically comprise of a cellular and biomaterial component. The latter takes on the role of a scaffold, to permit aggregation and organization of 3-dimensional extracellular matrix or tissue produced by the cells. However, it is clear that in areas of tissue engineering where tissue defects, voids, gaps or spaces need to be filled, proper *in situ* tissue formation will require sufficient spatial and temporal retention of the engineered construct at the repair site. Thus far, limited efforts have focused on this important design consideration required for the proper implantation of tissue engineered materials. Yet, the integration of the biomaterial with surrounding tissue is critical in maximizing long-term functionality and success. This can be achieved by one of 4 fixation methods: i) chemical, ii) mechanical, iii) biological or iv) some combination of i) through iii). Importantly, it is critical that the methodologies applied do not interfere with the tissue repair and regeneration processes. In this chapter, we review strategies for enhanced integration between biomaterial and native tissues, the ultimate goal being that *in situ* engineered tissue formation can be augmented. Emphasis is placed in the area of cartilage tissue engineering wherein strong integration between engineered cartilage and subchondral bone is required.

---

* Corresponding Author: Sharan Ramaswamy, Ph.D. Florida International University, Department of Biomedical Engineering, College of Engineering an Coputing, 10555 W. Flagler Street, EC 2612, Miami, FL 33174. Tel: (305)-348-2532; Fax: (305)-348-6954; Email: sramaswa@fiu.edu.

INTRODUCTION

Osteoarthritis (OA) is the most common joint disorder in humans. This disease involves degradation of articular cartilage, the sub chondral bone, the synovial capsule and membrane as well as the periarticular tissues [1]. According to the Centers for Disease and Control Prevention, 26.9 million individuals have been estimated to be affected by OA. The exact cause of this incurable disease is still unknown; however, articular cartilage degradation is prevalent in the progression of this disease.

Cartilage provides a cushioning mechanism to minimize impact to bone at joint locations, by completely covering its free surface. Conversely the degradation of articular cartilage leads to the exposure of bone and eventually, bone-to-bone contact which can cause considerable pain and stiffness in the joints. The extra cellular matrix of articular cartilage is composed of collagen –II and proteoglycans which are maintained and synthesized by chondrocyte cells. The tissue does not contain any vasculature hence attributing to the slow rate of growth and repair [2, 3]. Architecturally, cartilage is composed of 3 zones (Figure 1) and can be classified according to the orientation of collagen fibers within the tissue [4, 5]. The superficial layer has a tangential orientation with respect to the free surface, while the middle transitional layer is more random. Finally the deep zone is orientated orthogonally to the free surface.

Several methods have been implemented clinically to treat osteochondral defects but none have been able to produce a long term, durable solution [7, 8, 9]. Current clinically used strategies involve permanent modification of the damaged articular surface using abrasion arthoplastry or shaving of chondral bone to remove any loose bits or parts to ease the pain. Another popular technique in severe cases of OA makes use of grafting procedures such as asmosaicplasty [10, 11] or osteochondral allograft transplantation. In situations of complete cartilage loss and/or severe pain, total joint replacement is carried out such that an implant consisting of artificial materials is implanted. While these implants have demonstrated durability > 10 years, revisions are required and the procedure is expensive [12]. From a regenerative medicine standpoint, autologous chondrocyte transplantation (ACT) has been used to promote [13, 14] chondrogenesis. In addition, of keen interest in cartilage tissue engineering, several approaches involve leveraging bone marrow derived stem cell (BMSC) release from the subchondral bone which involve drilling [15] or microfracture [16] of the same with the hope that BMSCs will migrate in the defect area and promote cartilage repair. While these tissue engineering approaches seem promising, fundamentally, there is still a need to better integrate the biomaterial to the host to allow for sufficient repair and regeneration of new cartilage tissue.

Successful integration will require overcoming the following two major challenges: 1) the biomaterial scaffold that is used must initially be harmonized in the joint and the ensuing developing tissue must amalgamate with the existing native tissue that surrounds it; 2) the mechanically hostile environment of the joint poses further challenges and must thus be overcome by tailoring suitable interfacial mechanical properties after integration of the biomaterial to subchondral bone, so that *de novo* cartilage tissue can adequately form.

To our knowledge, different approaches of integration have been attempted but have not been systematically documented. To this end, the purpose of this book chapter is to provide a review of the attempted engineered cartilage to subchondral bone integration strategies and to offer our perspectives on the incorporation of a successful integration approach to further improve outcomes for cartilage tissue engineering.

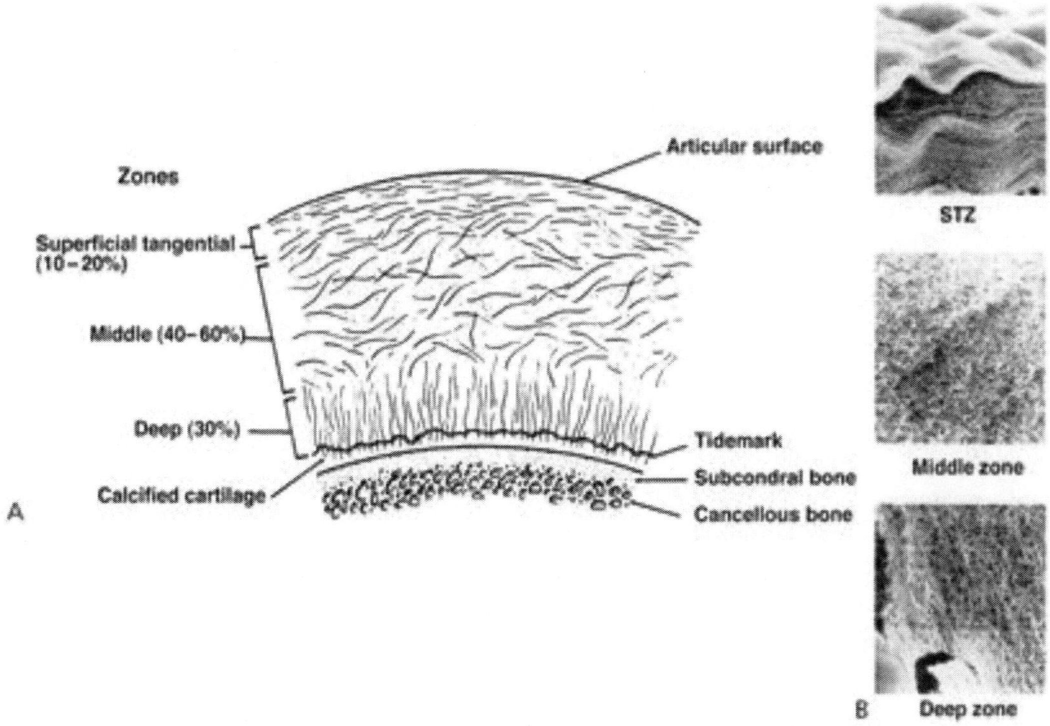

Reproduced with permission from: Van C. Mow, Rik Huiskes. Basic Orthopaedic Biomechanics and Mechano-Biology, 3rd Edition, Lippincott Williams & Wilkins, 2005 chapter 5, page 197, Fig 5-8 [6].

Figure 1. (A) Zonal structure of articular cartilage. The superficial zone contains collagen fibers aligned tangentially to the joint surface, whereas the intermediate zone has a more random orientation. Finally, the deep zone of cartilage is characterized by a perpendicular arrangement of collagen fibers. (B) Microscopic view of the three zones.

INTEGRATION APPROACHES

The process of integration of tissue engineered constructs can be broadly classified into three main categories.

a) Mechanical fixation
b) Chemical Fixation
c) Biological fixation

Mechanical Fixation

Sutures

Sutures have been extensively used to stabilize implants [17]. An ideal suture has characteristics of high tensile strength, excellent knot security, minimal tissue inflammation, and biocompatibility [18, 19]. Sutures are made of a variety of textile materials, including, Dexon, silk, nylon, polyethylene and Chitin [20, 21]. However sutures alone are not able to enable total fixation. Robertson *et al* [22] (1992) reported that the suturing process normally leads to the tearing of tissue when load is applied. Moreover premature suture breakage frequently occurs during application of a load. In some cases the suture threads actually pulled the implant away from the defect site resulting in dislodgement [22].

Screws

Screws have been utilized to anchor tissue engineered constructs to bone [23]. They act as a stabilizing device that helps to retain the construct in the defect. The main disadvantage of screws is that their protruding parts create regions of stress shielding and can alter bone material properties. In addition their protrusion may interfere with the seamless movement of joints at the articulating surfaces.

Cross Pins

To overcome the limitations of screws, cross pins were utilized so as to augment stability to the implant. Cross pins are much thinner than screws, thus allowing them to affix the tissue engineered constructs with the bone without tearing [24]. In this process, many research groups [25] used a pin fixture which degrades over time. Pins offer many advantages over screws, in that they prevent graft laceration and their heads are considerably softer than screws which reduces damage to the engineered constructs.

Staples

Robertson *et al.* [22] reported that use of staples is the most effective method of mechanical fixation of tissue engineered constructs to bone. However the most common mode of failure for staples was dislodgment after application of load. In addition, Shall *et al* [26] reported that staple breakage and pull-out were the two main reasons for fixation inadequacies.

Mechanical fixation manifests itself through a direct physical connection between implant and the bone. However the primary limitation in these approaches are that they form a sharp interface rather than a true integration between engineered to native tissue structures, which ultimately leads to implant loosening over a period of time. Considerable efforts have gone into determining novel ways of mechanically fixing implants to host tissue. However, true integration of engineered to native tissues such as bone cannot be achieved without biological growth and assimilation of these two entities. In addition, mechanical fixation alone may cause secondary tissue damage owing to new defect creation during processes such as suturing and stapling, which cannot be easily repaired [27].

Chemical Fixation

There are number of chemical-based methods that have been attempted to fix tissue engineered constructs to bone in animal models such as using tissue adhesives [28, 29], treating the construct with enzymes[41, 42] and blocking signaling pathways [45]. These methods have shown some success in terms of improving the mechanical stability and durability of the implant. Some of the approaches which have used a chemical approach are described below:

Tissue Adhesives

Tissue adhesives have been used for minor wound repair [28]. They have been implemented in place of sutures and staples for enabling improved outcomes for wound closure [29]. There are a variety of tissue adhesives that have been used clinically. The primary adhesives are described as follows:

Cyanoacrylates are a class of tissue adhesives that have been extensively used. There are many derivatives of cyanoacrylates such as N-butyl cyanoacrylate, 2-Octylcyanoacrylate, butyl-2-cyanoacrylate and N-butyl-2-cyanoacrylate [30, 31, 32]. These cyanoacrylates are commonly referred to as "superglues". They are liquid monomers which cause an exothermic reaction and form a flexible film at the tissue surface that will subsequently hold the two parts together [33]. Vote *et al* [34] (2000) reported that cyanoacrylate prevents the re-occurrence of epithelialization within the damaged area and the connective tissue. Cyanoacrylates also minimize infection because of their bacteriostatic properties which serve to inhibit the growth of gram positive organisms [31, 34].

Bioglue is another type of tissue adhesive that consists of albumin gluteraldehyde. It is composed of purified bovine serum albumin cross linked with gutaraldehyde. Herget *et al.* [35] (2001), suggested that this glue was suitable for sealing lung parenchyma and bronchial anastomosis defects. They furthermore reported that bioglue is durable and its adhesive capability is effective [35]. Several surgical procedures utilize fibrin glue. This tissue adhesive is composed of fibrinogen and thrombin linked together with a bond formed via the coagulation reaction [36]. Fibrin glue also consists of antifibrinolytic agents, aprotinin and calcium chloride in addition to the primary fibrinogen and thrombin components [37]. It has been successfully and actively used as a sealant for pulmonary airleaks [38]. In addition, although fibrin glue is found to be chemically less toxic then cyanoacrylates, they possess less adhesive strength properties [39], thus limiting their use in tissue integration processes.

Enzymatic Approaches

Enzymatic treatment approaches work on the principle that selective enzymatic treatment of tissue engineered constructs will promote integration to the host tissue and thereby provide better fixation. Many enzyme treatments have been attempted and have demonstrated improved integration outcomes. For example, Hunziker *et al* [41] suggested that the removal of proteoglycans from the surface would enhance the repair of cartilage. Their hypothesis was based on the fact that the cells that are responsible for repairing and regenerating the tissue in cartilage injury do not adhere to proteoglycan rich surfaces. Thus the removal of these proteoglycan molecules by the use of the enzyme, chondroitinase ABC would enhance the cartilage repair process. From their experiments, they were able to show that the repair cells adhered to the defect surface considerably better on removal of proteoglycan molecules; however, they were not able to show that the chondroitinase treatment actually enhanced the healing of the injured tissue. Lee *et al* [42] (2000) reported that chondroitinase treatment at various concentrations of the enzyme and exposure durations altered adhesion behavior of transplanted cells for the repair of the injured tissue. However, low concentrations of the enzyme (0.25 U/mL) for 15 minutes (and alternatively, 0.5 U/mL for 5 minutes) had no effect in improving adhesion of the cells. Bravenboer *et al*, [43] (2004), suggested that integration of engineered cartilage constructs can be enhanced by enzymatic treatment of collagenase and hyaluronidase. They further reported that treatment of these enzymes on the implanted tissue construct will increase the mechanical properties at the interface of engineered to host tissues. The same group had also previously showed that treatment with collagenase and hyaluronidase increased the cell density to the wound site [44]. Although there was significant increase in the integration strength of the implanted tissue model to native tissue using this dual enzymatic treatment, the reported values were still seven times lower than the actual strength of intact cartilage to subchondral bone [43]. Nonetheless, enzymatic treatments may offer a promising addition when used in conjunction with potential tissue integration strategies that utilize biological fixation methodologies.

Signaling Pathway /Biochemical Approaches

A blocking signaling pathway responsible for degrading tissue may indirectly promote the integration of the tissue engineered construct to native tissue. Recently Djouad *et al* [45] (2009) conducted experiments involving the blocking of the signaling pathway of extracellular regulated kinases (ERK). They found that disintegration of the extracellular matrix in nativecartilage is aggravated by the degradative signaling cascades which were initiated by proinflammatory cytokines. Thus blockage of extracellular regulated kinases 1 and 2 were thought to improve the integration processes between engineered to host cartilage. Elsewhere [46, 47], it was reported that interleukin 1-β(IL1-β) and tumor necrosis factor-α(TNF-α) are two important factors that are mainly responsible for causing the extracellular matrix (ECM) degradation and that they stimulate the formation of ERK 1/2, a protein molecule which is responsible for cartilage degradation. Djouad *et al* [45] were successful in blocking IL1-β and TNF-α signaling and subsequently were able to measure improved biochemical and mechanical properties at the interface of engineered cartilage and native tissue. However the *de novo* cartilage tissue as a whole had lower mechanical properties; for example, their control samples had a Young's modulus of 17.7- 5.5 kPa while their samples treated with IL1-b– and TNF-αrecorded values of 7.30-2.8 kPa and 11.5-1.7 kPa, respectively. These results suggests that integration processes using signaling pathway alteration approaches may improve at the cost of deteriorated engineered tissue bulk mechanical properties. Overall, a variety of chemical-based approaches have been used, but none of them have been capable of achieving durable integration with native tissues. Thus the demand for strong and durable integration of the engineered implant to native tissues [40] may only be possible with biological fixation measures.

Biological Fixation

Biological fixation approaches serve to enhance integration of tissues by true biological in-growth and amalgamation. Some of the techniques that are used have been categorized below.

Chondroitin Sulphate

The incapability of tissue adhesives to provide a stable and durable fixation over a period of time without causing any infection and maintaining the biocompatibility, particularly in the area of cartilage tissue engineering, led to integration studies involving multifunctional chondroitinsulphate [40]. Wang *et al* [40] (2007) reported that this material was chemically functionalized by two groups, methacrylate and aldehyde which can form a covalent bond to adhere to native cartilage on one side and the biomaterial scaffold on other side, thus forming a strong primer between the host and engineered tissues [40]. They also suggested that the material was bioactive as chondroitin sulphate is a principal component of cartilage ECMwhich would thereby serve to augment tissue repair.

Chitosan

Chitosan is a polysaccharide which is biocompatible, biodegradable, non toxic and allergen free. Chitosan has a hydrophilic surface which promotes both cell adhesion and proliferation [48]. Hoemann*et al* [49] (2006) indicated that Chitosan glycerol phosphate when mixed with blood in the cartilage defect spacenot only formed greatly improved integration with sub-chondral bone, but also resulted in enhanced hyaline cartilage tissue formation. However, chitosan has low strength [50], so it cannot be used alone in the fixation process.

Hydroxyapatite

Hydroxyapatite (or calcium phosphate) is a bioactive material found in the bone and teeth. A few research groups have been working on providing enhanced engineered cartilage to bone anchorage through the use of this bio active material [51, 52], including in our research laboratory. We are investigating if hydroxyapatite (HA) nanoparticles can be used to promote bone-in-growth at the bone-to-biomaterial interface of photopolymerizable hydrogels, thereby enhancing the retention of the tissue engineered constructs within the defective area for an appropriate healing period of around 4 weeks.

CURRENT AND FUTURE OUTLOOK

A summary of the different integration methods for engineered cartilage tissue constructs to bone, that were presented in this chapter is described in Table 1. Ultimately, providing a tissue engineered construct which can integrate to host cartilage and bone with augmented durability and stability remains a major challenge in cartilage tissue engineering. Our research efforts currently focus on the use of injectable polymers as a model system to assess integration processes and efficacy. In line with other investigators [53, 54], our motivation for such an approach is that injectable approaches have the ability to introduce a monomer solution to fill arbitrarily-shaped defects while concomitantly incorporating any secondary substance; polymerization (e.g. via UV exposure and photoinitiator inclusion) can then occur *in situ* [55].

Table 1. Summary of the advantages and disadvantages of different types of integration techniques [16-52]

Fixation type	Specific approaches	Advantages	Disadvantages
Mechanical Fixation	Sutures	- Extensively used for retaining construct with the bone - Inexpensive	- Premature suture Breakage - Tearing of Tissue engineered constructs
	Screws	- Helps in securing the constructs to bone	- Protruding heads leads to the tearing of tissue and affects the bone shear
	Cross Pin	- Thinner than screws and soft heads - prevents graft laceration	- Breakage
	Staples	- Most effective method for mechanical fixation	- Dislodgement of staples after the application of load. - Staple breakage - Staple pull out
Chemical Fixation	Tissue Adhesives	- Durable	- Some are not bio-compatible - They prevent the epithelialization within the damaged area and connective tissue - They may cause infection
	Enzymatic treatment	- promotes integration between the implant and the native tissue	- Alone this treatment cannot provide enough strength
	Biochemical Approach	- indirectly promotes the integration of the tissue engineered constructs with native tissue	- They deteriorate engineered tissue's bulk mechanical properties
Biological Fixation	Chondroitin Sulphate	- Bioactive material	Was not able to form a strong force of interaction between the host and the tissue engineered constructs
	Chitosan	- Biocompatible - Biodegradable - Allergy free - Non toxic	- Low Strength
	Hydroxyapatite	-Bioactive material - Helps in the formation of deep layer cartilage with bone marrow derived stem cells	- Ceramic, may break at a sudden load

Our immediate interests are with the inclusion of hydroxyapatite in a photopolymerizable hydrogel system, such as an osteoinductive material, to improve integration between subchondral bone and tissue engineered cartilage. Khanarian *et al* [51] (2012) have already utilized alginate gels with hydroxyapatite and demonstrated improved outcomes for integration between engineered and host cartilage, through the formation of a calcified cartilage-like matrix interface. We are currently studying this process in photopolymerizable hydrogels for injectable tissue engineering [56] and are *en route* to evaluating the efficacy of integration between the cultured engineered cartilage and bone.

CONCLUSION

Current paradigms in cartilage tissue engineering do not sufficiently address the issue of integrating engineered construct to native cartilage and bone. The different approaches relating to this critical area, which are mechanically, chemically or biologically driven, have been discussed in this review. We expect that biological processes in conjunction with a mechanical and/or chemical approach are ultimately going to be essential in permitting a gradual transition of tissue properties surrounding the interface and thus, true integration can be achieved. Such a gradual transition is thought to be beneficial in promoting long term retention of the engineered construct within the defect space [57]. Renewed interest in this area by several research groups , including our own laboratories [40, 50, 52, 57, 58] may provide the necessary breakthrough in enhancing integration outcomes and in doing so, moving from benchtop and *in vivo* studies to functional assessment in clinical trials, over the next decade.

REFERENCES

[1] Y. Henrotin, B. Kurz and T. Aigner. Oxygen and reactive oxygen species in cartilage degradation: friends or foes. *OsteoArthritis and Cartilage* (2005) 13, 643-54.

[2] Hunter W: On the structure and diseases of articulating cartilages. *Trans R Soc Lond* 1743, 42B:514-521.

[3] Newman AP: Articular cartilage repair. *Am J Sports Med* 1998, 26:309-324.

[4] Blanka Sharma and Jennifer H. Elisseeff. Engineering Structurally Organized Cartilage and Bone Tissues. *Annals Of Biomedical Engineering*. Volume 32, Number 1, 148-59.

[5] T.-K. Kim, B. Sharma, C. G. Williams, M. A. Ruffner, A. Malik, E. G. McFarland and J. H. Elisseeff. Experimental Model for Cartilage Tissue Engineering to Regenerate the Zonal Organization of Articular Cartilage. *OsteoArthritis and Cartilage* (2003) 11, 653–64.

[6] Van C. Mow, Rik Huiskes. Basic Orthopaedic Biomechanics and Mechano-Biology,3rd Edition, Lippincott Williams & Wilkins, 2005..

[7] Spalazzi J.P., Doty S.B., Moffat K.L., Levine W.N., and Lu H.H. Development of controlled matrix heterogeneity on a triphasic scaffold for orthopedic interface tissue engineering. *Tissue Eng*. 2006 Dec;12(12). P. 3497-08.

[8] A.A. van Valburg, P.M. van Roermund, A.C.A. Marijnissen, J. van Melkebeek, J. Lammens, A.J. Verbout and et al. Joint distraction in treatment of osteoarthritis: a two-year follow-up of the ankle. *Osteoarthritis and Cartilage.* Volume 7, Issue 5, September 1999, Pages 474–79.

[9] J. A. Buckwalter, H. J. Mankin. Articular cartilage repair and transplantation. *Arthritis Rheum*, 1998. Volume 41, Issue 8, pages 1331–42.

[10] Hangody L, Kish G, Karpati Z, Udvarhelyi I, Szigeti I, Bely M: Mosaicplasty for the treatment of articular cartilage defects: application in clinical practice. *Orthopedics* 1998, 21:751-756.

[11] Feczko P, Hangody L, Varga J, Bartha L, Dioszegi Z, Bodo G, Kendik Z, Modis L: Experimental results of donor site filling for autologous osteochondral mosaicplasty. *Arthroscopy* 2003, 19:755-761.

[12] MD Stephen L. Brotherton, James R. Roberson, J. Robin de Andrade, Lamar L. Fleming. Staged versus simultaneous bilateral total knee replacement. *The Journal of Arthroplasty* Volume 1, Issue 4, 1986, Pages 221–28.

[13] Brittberg M, Lindahl A, Nilsson A, Ohlsson C, Isaksson O, Peterson L: Treatment of deep cartilage defects in the knee with autologous chondrocyte transplantation. *N Engl J Med* 1994, 331:889-895.

[14] Brittberg M, Peterson L, Sjogren-Jansson E, Tallheden T, Lindahl.A. Articular cartilage engineering with autologous chondrocyte transplantation. A review of recent developments. *J Bone Joint Surg Am* 2003, Suppl 3:109-115.

[15] Beiser IH, Kanat IO. Subchondral bone drilling: a treatment for cartilage defects. *J Foot Surg.* 1990 Nov-Dec;29(6):595-01.

[16] Frisbie DD, Trotter GW, Powers BE, Rodkey WG, Steadman JR, Howard RD and et al. Arthroscopic Subchondral Bone Plate Microfracture Technique Augments Healing of Large Chondral Defects in the Radial Carpal Bone and Medial Femoral Condyle of Horses. *Veterinary Surgery*, 1999. Volume 28, Issue 4, pages 242–55.

[17] Matthews LS, Lawrence SJ, Yahiro MA, Sinclair MR. Fixation strengths of patellar tendon-bone grafts. *Arthroscopy.* 1993;9(1):76-81.

[18] John B. Herrmann, Richard J. Kelly, G*eorge A. Higgins. Polyglycolic Acid Sutures. AMA Arch Surg.* Vol. 100 No. 4 *1970.* P: 486-490.

[19] Field JR, Stanley RM. Suture characteristics following incubation in synovial fluid or phosphate buffered saline. *Injury.* 2004 Mar;35(3):243-8.

[20] Mouzas GL, Yeadon A. Does the choice of suture material affect the incidence of wound infection? A comparison of dexon (polyglycolic acid) sutures with other commonly used sutures in an accident and emergency department. *Br J Surg.* 1975 Dec;62(12):952-5.

[21] Nakajima M, Atsumi K, Kifune K, Miura K, Kanamaru H. Chitin is an effective material for sutures. *Jpn J Surg.* 1986 Nov;16(6):418-24.

[22] Daniel B. Robertson, Dale M. Daniel, Edmund Biden. Soft tissue fixation of bone. *Am J Sports Med.* 1986;14(5):398-40.

[23] Stengel D, Casper D, Bauwens K, Ekkernkamp A, Wich M. Bioresorbable pins and interference screws for fixation of hamstring tendon grafts in anterior cruciate ligament reconstruction surgery: a randomized controlled trial. *Am J Sports Med.* 2009 Sep;37 (9):1692-8. Epub 2009 May 29.

[24] Zantop T, Welbers B, Weimann A, Rümmler M, Hedderich J, Musahl V, Petersen W. Biomechanical evaluation of a new cross-pin technique for the fixation of different sized bone-patellar tendon-bone grafts. *Knee Surg Sports Traumatol Arthrosc*. 2004 Nov;12(6):520-7. Epub 2004 Jun 19.

[25] Zantop T, Weimann A, Rümmler M, Hassenpflug J, Petersen W. Initial fixation strength of two bioabsorbable pins for the fixation of hamstring grafts compared to interference screw fixation: single cycle and cyclic loading. *Am J Sports Med*. 2004 Apr-May;32(3):641-9.

[26] Shall LM, Cawley PW. Soft Tissue Reconstruction in the Shoulder. *Am J Sports Med*. 1994 Sep-Oct;22(5):715-8.

[27] Hunziker, E. B. Articular cartilage repair: Basic science and clinical progress. A review of the current status and prospects. *Osteoarthritis Cartilage*. 2002 Jun;10(6):432-63.

[28] Bruns TB, Worthington JM. Using tissue adhesive for wound repair: a practical guide to dermabond. *Am Fam Physician*. 2000 Mar 1;61(5):1383-8.

[29] Coulthard P, Esposito M, Worthington HV, van der Elst M, van Waes OJ, Darcey J. Tissue adhesives for closure of surgical incisions. *Cochrane Database Syst Rev*. 2010 May 12;(5):CD004287.

[30] Leahey AB, Gottsch JD, Stark WJ. Clinical experience with N-butyl cyanoacrylate (Nexacryl) tissue adhesive. *Ophthalmology*. 1993 Feb;100(2):173-80.

[31] Singer AJ, Hollander JE, Valentine SM, Turque TW, McCuskey CF, Quinn JV. Prospective, randomized, controlled trial of tissue adhesive (2-octylcyanoacrylate) vs standard wound closure techniques for laceration repair. Stony Brook Octylcyanoacrylate Study Group. *Acad Emerg Med*. 1998 Feb;5(2):94-9.

[32] Woodward SC, Herrmann JB, Cameron JI, Brandes G, Pulaski EJ, Leonard F. Histotoxicity of cyanoacrylate tissue adhesive in the rat. *Ann Surg*. 1965 Jul;162:113-22.

[33] Singer AJ. A review of the literature on octylcyanoacrylate tissue adhesive. *Am J Surg*. 2004 Feb;187(2):238-48.

[34] Vote BJ, Elder MJ. Cyanoacrylate glue for corneal perforations: a description of a surgical technique and a review of the literature. *Clin Experiment Ophthalmol*. 2000 Dec;28(6):437-42.

[35] Herget GW, Kassa M, Riede UN, Lu Y, Brethner L, Hasse J. Experimental use of an albumin-glutaraldehyde tissue adhesive for sealing pulmonary parenchyma and bronchial anastomoses. *Eur J Cardiothorac Surg*. 2001 Jan;19(1):4-9.

[36] Mizrahi B, Weldon C, Kohane DS. Tissue Adhesives as Active Implants. Stud Mechanobiol *Tissue Eng Biomater* (2011) 8: 39–56.

[37] Dunn CJ, Goa KL. Fibrin sealant: a review of its use in surgery and endoscopy. *Drugs*. 1999 Nov;58(5):863-86.

[38] Kjaergard HK, Weis-Fogh US, Sørensen H, Thiis J, Rygg I. Autologous fibrin glue--preparation and clinical use in thoracic surgery. *Eur J Cardiothorac Surg*. 1992;6 (1):52-4; discussion 54.

[39] Lauto AA. Adhesive biomaterials for tissue reconstruction. Journal of chemical technology and biotechnology (1986). 2008 -04;83(4):464-72.

[40] Wang DDA. Multifunctional chondroitin sulphate for cartilage tissue–biomaterial integration. *Nature materials*. 2007 -05;6(5):385-92.

[41] Hunziker EB, Kapfinger E. Removal of proteoglycans from the surface of defects in articular cartilage transiently enhances coverage by repair cells. *J Bone Joint Surg Br.* 1998 Jan;80(1):144-50.

[42] Lee MC, Sung KL, Kurtis MS, Akeson WH, Sah RL. Adhesive force of chondrocytes to cartilage. Effects of chondroitinase ABC. *Clin Orthop Relat Res.* 2000 Jan;(370):286-94.

[43] van de Breevaart Bravenboer J, In der Maur CD, Bos PK, Feenstra L, Verhaar JA, Weinans H, van Osch GJ. Improved cartilage integration and interfacial strength after enzymatic treatment in a cartilage transplantation model. *Arthritis Res Ther.* 2004;6(5):R469-76. Epub 2004 Aug 6.

[44] Bos PK, DeGroot J, Budde M, Verhaar JA, van Osch GJ: Specific enzymatic treatment of bovine and human articular cartilage: implications for integrative cartilage repair. *Arthritis Rheum* 2002, 46:976-985.

[45] Djouad F, Rackwitz L, Song Y, Janjanin S, Tuan RS. ERK1/2 Activation Induced by Inflammatory Cytokines Compromises Effective Host Tissue Integration of Engineered Cartilage. *Tissue Eng Part A.* 2009 Oct;15(10):2825-35.

[46] Goldring MB, Berenbaum F. The regulation of chondrocyte function by proinflammatory mediators: prostaglandins and nitric oxide. *Clin Orthop Relat Res.* 2004 Oct;(427 Suppl):S37-46.

[47] Lianxu C, Hongti J, Changlong Y. NF-kappaBp65-specific siRNA inhibits expression of genes of COX-2, NOS-2 and MMP-9 in rat IL-1beta-induced and TNF-alpha-induced chondrocytes. *Osteoarthritis Cartilage.* 2006 Apr;14(4):367-76. Epub 2005 Dec 22.

[48] Li Z, Ramay HR, Hauch KD, Xiao D, Zhang M. Chitosan–alginate hybrid scaffolds for bone tissue engineering. *Biomaterials.* 2005 Jun;26(18):3919-28.

[49] Hoemann CD, Sun J, McKee MD, Chevrier A, Rossomacha E, Rivard GE, Hurtig M, Buschmann MD. Chitosan–glycerol phosphate/blood implants elicit hyaline cartilage repair integrated with porous subchondral bone in microdrilled rabbit defects. *Osteoarthritis Cartilage.* 2007 Jan;15(1):78-89. Epub 2006 Aug 8.

[50] Madihally SV, Matthew HW. Porous chitosan scaffolds for tissue engineering. *Biomaterials.* 1999 Jun;20(12):1133-42.

[51] Khanarian NT, Jiang J, Wan LQ, Mow VC, Lu HH. A Hydrogel-Mineral Composite Scaffold for Osteochondral Interface Tissue Engineering. *Tissue Eng Part A.* 2012 Mar;18(5-6):533-45. Epub 2011 Nov 8.

[52] Dua R, Centeno J, Ramaswamy S. Enhanced Anchorage of Tissue Engineered Cartilage Constructs using Hydoxyapatite nanoparticles. Conference presentation at BMES Annual Meeting in Hartford, CT, October 12-15, 2011.

[53] Ramaswamy S, Gurkan I, Sharma B, Cascio B, Fishbein KW, Spencer RG. Assessment of tissue repair in full thickness chondral defects in the rabbit using magnetic resonance imaging transverse relaxation measurements. *J Biomed Mater Res B Appl Biomater.* 2008 Aug;86(2):375-80.

[54] Toh WS, Lee EH, Guo XM, Chan JK, Yeow CH, Choo AB, Cao T. Cartilage repair using hyaluronan hydrogel-encapsulated human embryonic stem cell-derived chondrogenic cells. *Biomaterials.* 2010 Sep;31(27):6968-80. Epub 2010 Jun 17.

[55] Gutowska A, Jeong B, Jasionowski M. Injectable Gels for Tissue Engineering. *Anat Rec*. 2001 Aug 1;263(4):342-9.

[56] Tan H, Chu CR, Payne KA, Marra KG. Injectable In Situ Forming Biodegradable Chitosan-Hyaluronic acid Based Hydrogels for Cartilage Tissue Engineering. *Biomaterials*. 2009 May;30(13):2499-506. Epub 2009 Jan 23.

[57] Ramaswamy S, Wang DA, Fishbein KW, Elisseeff JH, Spencer RG. An analysis of the integration between articular cartilage and nondegradable hydrogel using magnetic resonance imaging. *J Biomed Mater Res B Appl Biomater*. 2006 Apr;77(1):144-8.

[58] Dua R, Centeno J, Ramaswamy S. Novel design to integrate tissue engineered cartilage to native bone using hydroxyapatite nanoparticles. *IEEE conference publications*, 2011: 418 - 420.

In: Tissue Engineering
Editors: T. Ishikawa and M. Yamamoto
ISBN 978-1-62257-062-1

Chapter 5

ADIPOSE-DERIVED STEM CELLS AND THEIR APPLICATION IN TISSUE ENGINEERING

Natsuko Kakudo**, Satoshi Kushida, Tsunetaka Ogura, Kenji Suzuki and Kenji Kusumoto
Department of Plastic and Reconstructive Surgery,
Kansai Medical University,
Japan, Fumizono, Moriguchi, Osaka, Japan

ABSTRACT

Regenerative medicine is emerging as a rapidly evolving field of research and therapeutics, and stem cells hold great promise for future translational research and widespread clinical application. Adipose-derived stem cells (ASCs) are as capable as bone marrow-derived stem cells (BMSCs) of differentiating into multiple lineages, such as bone, cartilage, tendons, nerves and fat, when cultivated under lineage-specific conditions.

Furthermore, ASCs appear to have comparable properties to BMSCs as well as favorable angiogenic and anti-inflammatory secretion profiles. As human adipose tissue can be obtained easily in large quantities under local anesthesia with little patient discomfort, it may thus provide an alternative source of stem cells for mesenchymal tissue regeneration and engineering.

In this chapter, we describe the methods used in our laboratory and in the literature for the isolation and expansion of ASCs. In addition, we present the potential application of ASCs in tissue engineering, particularly in the regeneration of fat, bone and cartilage and comment on the progress of this approach and future trends in the field. The fast expanding field of basic and clinical research into ASC biology, pathophysiology, and potential therapeutic applications promises further exciting discoveries in the near future.

* Tel: +81-6-6992-1001; Fax: +81-6-6997-0628; E-mail: kakudon@takii.kmu.ac.jp.

INTRODUCTION

Adipose tissue has recently been confirmed as another source of multipotent adult stem cells, entitled adipose-derived stem cells (ASCs). Multilineage stem cells derived from the stroma of adipose tissue were first described by Zuk et al., who isolated these cells from lipoaspirate [1].

They reported that ASCs differentiate in vitro into adipogenic, chondrogenic, myogenic, and osteogenic cells in the presence of lineage-specific induction factors, demonstrating support for the hypothesis that human lipoaspirate contains multipotent cells and may represent an alternative stem cell source to bone marrow-derived MSCs (BMSCs).

Furthermore, one gram of adipose tissue has been shown to yield approximately 5×10^3 stem cells, which is 500-fold greater than the number of stem cells in one gram of bone marrow [2].

Human adipose tissue can be obtained repeatedly in large quantities under local anesthesia with minimal patient discomfort, and the convenient isolation, extensive in vitro proliferative capacity and differentiation potential of autologous ASCs make them a promising source of human stem cells for regenerative medicine.

In this chapter, we describe the methods used in our laboratory and in the literature for the isolation and expansion of ASCs. In addition, we present the potential application of ASCs in tissue engineering; particularly for the regeneration of fat, bone and cartilage, and comment on progress in this approach and future trends in the field.

ISOLATION OF ADIPOSE-DERIVED STEM CELLS

In previous studies, we established a standard protocol for the isolation of ASCs from adipose tissue using collagenase enzymatic digestion [3]. This protocol was partially modified from the method reported by Zuk et al. [1]. Briefly, human adipose tissue is washed three times with 20 ml of phosphate buffered saline (PBS), cut into small pieces, and the extracellular matrix digested using 0.1% collagenase solution with shaking at 37°C for 40 min. After the addition of Dulbecco's modified Eagle's medium (DMEM) containing 10% fetal bovine serum (FBS) and 1% penicillin, the pieces are centrifuged at 1,500 rpm for 3 min and washed 3 times using DMEM-10% FBS.

Since prolonged digestion with collagenase and insufficient washing reduce cell viability and influence proliferation and differentiation, these processes are carefully performed. The supernatant is discarded and the pellet is resuspended and filtered through a 100-μm cell strainer to remove undigested tissue fragments. The cells are pelleted and then resuspended in the above cell culture medium. The adhered ASCs are maintained in this medium until passage 3, and nearly all cells form fibroblast-like cells.

THE PHENOTYPE OF ADIPOSE-DERIVED STEM CELLS

The cell surface phenotype of human ASCs is similar to that of BMSCs. For example, CD105, CD166 (ALCAM), and STRO-1 are common markers used to identify cells with multilineage differentiation potential and are consistently expressed on both ASCs and

BMSCs [4]. In addition, CD117 (the stem cell factor receptor) is expressed on an array of totipotent or pluripotent stem cells, including embryonic stem cells, hematopoietic stem cells, BMSCs, and ASCs [4].

ASCs and BMSCs also express CD9, CD10, CD13, CD29, CD44, CD49e, CD54, CD55, CD59, CD90, CD105, CD117 and CD146. Surface markers absent on both cell types include CD34 and CD45 [4]. However, it has been reported that, although fresh ASCs are CD105(–) (as well as CD31(–), CD34(+), CD45(–), CD90(+), and CD146(–)), these cells strongly express CD105 in culture [5]. Causes of this inconsistency include changes in ASC markers due to culture conditions and the number of passages.

The expression of CD34, CD90, CD73, CD44, and CD29 alters during ASC passages [6]. Mitchell's study reported that although expression of the stem cell-associated marker CD34 peaked in early-passage ASCs, it remained present at reduced levels throughout the culture period. Despite several studies to establish markers of the ASC phenotype, controversy regarding phenotype characterization persists.

Cytokine Profile Secreted by Adipose-Derived Stem Cells

ASCs secrete many types of cytokines that are related to angiogenesis and proliferation. Rehmann et al [7]. reported that ASCs secrete 1203+/-254 pg of vascular endothelial growth factor (VEGF), 12280+/-2944 pg of hepatocyte growth factor (HGF) and 1247+/-346 pg of transforming growth factor-beta per (TGF-β) per 10^6 cells over 72 h. Furthermore, when ASCs were cultured under hypoxic conditions, VEGF secretion increased 5-fold to 5980+/-1066 pg/10^6 cells.

This result is consistent with other studies, implying that hypoxia is a significant stimulus for ASC-mediated angiogenesis. A study by Kim et al. reported the following cytokine levels in the media of ASCs that were cultivated for 72 h: platelet-derived growth factor (PDGF; 44.41 ± 2.56 pg mL^{-1}), placenta-derived growth factor (PIGF; 37.87 ± 1.69 pg mL^{-1}), bFGF (131.35 ± 30.31 pg mL^{-1}), keratinocyte growth factor (KGF; 86.28 ± 20.33 pg mL^{-1}), TGF-β1 and –β2 (103.33 ± 1.70 and 75.42 ±95.98 pg mL^{-1}, respectively), HGF (670.94 ± 86.92 pg mL^{-1}), and VEGF (809.53 ± 95.98 pg mL^{-1}) [8]. In addition, large amounts of collagen and fibronectin were detected. These data suggest that factors secreted by ASCs may promote wound healing and thus constitute potential promoters of angiogenesis and proliferation.

Adipose-Derived Stem Cells for Bone Tissue Engeeneering

Several studies regarding the use of ASCs in the field of bone tissue engineering have already been reported. The first was that of Cowan et al. who investigated the in vivo osteogenic ability of ASCs, BMCs, calvarial-derived osteoblasts and dura mater cells to heal critical-size mouse calvarial defects. Implanted, apatite-coated, PLGA scaffolds seeded with ASCs or BMCs produced significant intramembranous bone formation by 2 weeks, and areas of complete bony bridging by 12 weeks, as shown by X-ray analysis, histology and live micromolecular imaging [9].

Subsequently, Peterson et al. conducted a study in which they implanted collagen-ceramic carriers seeded with ASCs transfected with the BMP-2 gene into nude mice for treatment of a critical-size femoral defect [10]. These carriers were able to deliver the cells to the defect area, the ASCs formed new bone due to the effect of BMP-2 and the femoral defects were thus healed 8 weeks after implantation [10].

Shen et al. demonstrated the fate of ASCs seeded onto three-dimensional sintered microsphere matrices (SMM). They prepared sintered microsphere matrices of poly(lactide-co-glycolide) [85:15] using a solvent evaporation technique. These matrices were then mechanically sieved and fabricated in metal molds by heating. ASCs from adult Sprague Dawley rats were then seeded onto the bioengineered SMM and cultured in media with growth and differentiation factor-5 (GDF-5). The results of this study confirmed that ASCs treated with GDF-5 are not only capable of adhering to a bioengineered scaffold, but also remain viable and demonstrate the ability to migrate, proliferate, and subsequently undergo osteogenic differentiation [11].

Hattori et al. compared the osteogenic differentiation potential of human BSCs and ASCs using β-Tricalcium Phosphate (β-TCP) scaffolds. The cells were cultured in vitro on the scaffold constructs with osteogenic medium for 2 weeks and were then implanted into the backs of nude mice for 8 weeks. The results showed that the ability of ASCs in a β-TCP disk to differentiate into osteoblasts and to synthesize bone is similar to that of BSCs [12].

We have also reported the osteogenic differentiation of ASCs on honeycomb collagen scaffolds and demonstrated that these ASCs differentiate into osteoblast-like cells, are able to produce matrix, and express the characteristic genes of osteoblasts. The ASC-loaded honeycomb collagen scaffolds that were cultured for 14 days were subcutaneously transplanted into nude mice and excised after 8 weeks. Bone formation in vivo was examined using HE stain, von Kossa stain and osteocalcin immunostaining. We concluded that that this carrier is a suitable scaffold for ASCs and will be useful as a three-dimensional bone tissue engineering scaffold in vitro and in vivo [13].

ADIPOSE-DERIVED STEM CELLS FOR CARTILAGE TISSUE ENGINEERING

Several studies have demonstrated the potential of ASCs in several different cartilage tissue engineering approaches, showing the chondrogenic differentiation potential when these cells are cultured using the typical aggregate culture technique, as pellet culture, and when they are seeded in material to be further cultured in vitro and/or implanted in different animals [14].

Award et al. reported one of the first studies on cartilage tissue engineering using ASCs. They compared several different scaffolds based on agarose, alginate, and gelatin as support materials for human ASC adhesion, proliferation, and chondrogenic differentiation. It was concluded that these cells could differentiate into chondrocytes when cultured in any of the studied scaffolds, although differences were found in the proliferation rate, indicating that the most appropriate scaffold material for cell growth was gelatin [15].

Betre et al. showed ASC chondrogenic differentiation using elastin-like polypeptide as support matrices and demonstrated that the ASCs can grow and express the chondrogenic phenotype, namely, the expression of collagen II, aggrecan, and sox9, even when cultured without the supplemented media usually used to induce chondrogenic differentiation [16].

Wei et al. studied the use of an injectable scaffold based on chondroitin sulfate for the ASC chondrogenic differentiation. The results showed good cell viability and growth and also that the ASCs differentiate into chondrocyte-like cells producing GAG [14].

Dragoo et al. studied in vivo chondrogenic regeneration by implanting fibrin glue scaffolds seeded with ASCs in rabbit for 8 weeks. At the end of the implantation period, the ASCs expressed aggrecan and collagen type II, the cells synthesized GAGs, and new cartilage formation was observed [17].

Other researchers reported that PLGA is a good scaffold used in cartilage tissue engineering based on differentiated ASCs [18] [19] [20]. Jin et al. reported that chondrogenic differentiation of human ASCs induced by transduction with replication-deficient adenovirus carrying human transforming growth factor beta2 (Ad5-hTGF beta2) was demonstrated by RT-PCR, immunohistochemistry staining, and biochemical and western blot analyses. To evaluate whether the in vitro differentiated ASCs could retain their chondrocytic phenotype and produce neo-cartilage in vivo, predifferentiated ASCs were seeded in different scaffolds and implanted in subcutaneous pockets on the dorsum of nude mice. After 4 and 12 weeks of culture in vivo, specimens were harvested and examined by histological and immunohistochemical analyses; cartilage-like tissue formation was only found in alginate gel and PLGA/alginate compound groups. In the PLGA group, fibrous tissues and angiogenesis ingrowth were observed. These findings demonstrated that adenovirus-mediated hTGF beta2 gene transfer could induce ASCs into a chondrogenic lineage in vitro; however, this predifferentiation did not guarantee ectopic cartilage formation in vivo unless appropriate three-dimensional scaffolds were used as the cell-carrying vehicles [21].

In another approach, ASCs were encapsulated in aliginate microbeads together with a chimeric RGD-protein. The analyzed parameters of gene expression (Sox9, Aggrecan, fibronectin, and collagen II) included chondrogenic differentiation after TGFβ3-dependent induction. Increased expression of chondrogenic genes was observed, which was dependent on β1 integrin-mediated signaling [22].

Cheng et al. worked without stimulation with exogenous growth factors when ASCs were seeded on scaffolds derived from porcine articular cartilage. After a cultivation period between 4 and 6 weeks, expression of chondrogenic genes was observed [23].

Hildner et al. determined that cocultivation of human articular chondrocytes together with ASCs resulted in increased collagen type IX expression, indicating a long-term stability of cartilage [24].

Adipose-Derived Stem Cells for Adipose Tissue Engineering

Patrick et al. reported that poly(lactic-co-glycolic) acid (PLGA) polymer disks with 90% porosity, 2.5 mm thickness, 12 mm diameter, and pore size in the range of 135-633 microns were fabricated and seeded with preadipocytes of rats. They demonstrated the potential of using primary preadipocytes as a cell source in cell-seeded polymer scaffolds for tissue engineering applications [25].

Lee et al. demonstrated that adipogenic differentiated rat ASCs were seeded on polyglycolic acid (PGA) fibers and implanted subcutaneously onto the head of rats. Intracellular lipid vacuoles could be demonstrated histologically in the grafts explanted after 4 and 8 weeks [26].

Von Heimburg et al. reported that human preadipocytes were seeded on different biodegradable carriers like HYAFF 11 sponge and collagen sponges [27]. The HYAFF 11 sponges turned out to have a higher cell density than the collagen matrices used [27]. Collagen sponges were also used by Huss et al. for coculture assay with human mammary epithelial cells and preadipocytes. They found lipid-containing cells clustered around ductal structures [28].

Hong et al. reported that gelatin sponges seeded with human ASCs have soft tissue-like mechanical properties, which are generally preferable compared with more rigid structures [29].

Halbleib et al. reported that HA-based scaffold has also been used as a scaffold for adipocyte precursor cells. They concluded that hyaluronic acid-based scaffolds appear to be a suitable three-dimensional carrier for the culture and in vitro differentiation of human adipocyte precursor cells [30].

Mizuno et al. reported in vivo adipose tissue regeneration by adipose-derived stromal cells isolated from GFP transgenic mice with fibrin glue [31]. They demonstrated that induced ASCs were merged with fibrin glue, and the mixture was injected subcutaneously into the dorsum of athymic mice, and both hematoxylin and eosin and oil red O staining revealed that cells containing small lipid droplets had been regenerated histologically [31].

Vallée et al. completely renounced the substitution of exogenous matrix materials, resulting in three-dimensional cell sheets of adipogenic stimulated cells [32]. They reported the successful in vitro production of human adipose substitutes featuring an increased surface area (>30 cm^2), and concluded that customized autologous reconstructed adipose tissues could be produced in the future to repair a wide range of soft-tissue defects [32].

Clinical Applications of Adipose-Derived Stem Cells in Humans and Future Perspective

A comparison between mesenchymal stem cells, BMSCs and ASCs, demonstrated more frequent clinical applications of BMSCs. However, the clinical applications of ASCs in humans have gradually increased since 2002 because of their higher availability and encouraging results from basic studies on their differentiation. For routine clinical applications in humans, stem cells are collected from adipose tissue to be implanted into an affected area without culturing. For plastic surgery, such as breast augmentation and reconstruction, as well as treatment of facial fat atrophy, the mixed transplantation of autologous ASCs and free adipose tissue has successfully increased tissue [33] [34]. In some cases, autologous ASCs were seeded onto a hyaluronan-based scaffold, and a cultured device was subcutaneously implanted [35]. Preclinical studies have successfully been conducted on ASCs, demonstrating improved blood flow in ischemic tissues of the skin and extremities [36], nerve regeneration for spinal cord injury [37], and improved liver functions [38].

REFERENCES

[1] Zuk, P. A., Zhu, M., Mizuno, H., Huang, J., Futrell, J. W., Katz, A. J., Benhaim, P., Lorenz, H. P., and Hedrick, M. H. (2001) Multilineage cells from human adipose tissue: implications for cell-based therapies, *Tissue engineering 7*, 211-228.

[2] Fraser, J. K., Wulur, I., Alfonso, Z., and Hedrick, M. H. (2006) Fat tissue: an underappreciated source of stem cells for biotechnology, *Trends in biotechnology 24*, 150-154.

[3] Kakudo, N., Shimotsuma, A., and Kusumoto, K. (2007) Fibroblast growth factor-2 stimulates adipogenic differentiation of human adipose-derived stem cells, *Biochemical and biophysical research communications 359*, 239-244.

[4] Strem, B. M., Hicok, K. C., Zhu, M., Wulur, I., Alfonso, Z., Schreiber, R. E., Fraser, J. K., and Hedrick, M. H. (2005) Multipotential differentiation of adipose tissue-derived stem cells, *The Keio journal of medicine 54*, 132-141.

[5] Yoshimura, K., Shigeura, T., Matsumoto, D., Sato, T., Takaki, Y., Aiba-Kojima, E., Sato, K., Inoue, K., Nagase, T., Koshima, I., and Gonda, K. (2006) Characterization of freshly isolated and cultured cells derived from the fatty and fluid portions of liposuction aspirates, *Journal of cellular physiology 208*, 64-76.

[6] Mitchell, J. B., McIntosh, K., Zvonic, S., Garrett, S., Floyd, Z. E., Kloster, A., Di Halvorsen, Y., Storms, R. W., Goh, B., Kilroy, G., Wu, X., and Gimble, J. M. (2006) Immunophenotype of human adipose-derived cells: temporal changes in stromal-associated and stem cell-associated markers, *Stem Cells 24*, 376-385.

[7] Rehman, J., Traktuev, D., Li, J., Merfeld-Clauss, S., Temm-Grove, C. J., Bovenkerk, J. E., Pell, C. L., Johnstone, B. H., Considine, R. V., and March, K. L. (2004) Secretion of angiogenic and antiapoptotic factors by human adipose stromal cells, *Circulation 109*, 1292-1298.

[8] Kim, W. S., Park, B. S., Kim, H. K., Park, J. S., Kim, K. J., Choi, J. S., Chung, S. J., Kim, D. D., and Sung, J. H. (2008) Evidence supporting antioxidant action of adipose-derived stem cells: protection of human dermal fibroblasts from oxidative stress, *Journal of dermatological science 49*, 133-142.

[9] Cowan, C. M., Shi, Y. Y., Aalami, O. O., Chou, Y. F., Mari, C., Thomas, R., Quarto, N., Contag, C. H., Wu, B., and Longaker, M. T. (2004) Adipose-derived adult stromal cells heal critical-size mouse calvarial defects, *Nature biotechnology 22*, 560-567.

[10] Peterson, B., Zhang, J., Iglesias, R., Kabo, M., Hedrick, M., Benhaim, P., and Lieberman, J. R. (2005) Healing of critically sized femoral defects, using genetically modified mesenchymal stem cells from human adipose tissue, *Tissue engineering 11*, 120-129.

[11] Shen, F. H., Zeng, Q., Lv, Q., Choi, L., Balian, G., Li, X., and Laurencin, C. T. (2006) Osteogenic differentiation of adipose-derived stromal cells treated with GDF-5 cultured on a novel three-dimensional sintered microsphere matrix, *The spine journal : official journal of the North American Spine Society 6*, 615-623.

[12] Hattori, H., Masuoka, K., Sato, M., Ishihara, M., Asazuma, T., Takase, B., Kikuchi, M., and Nemoto, K. (2006) Bone formation using human adipose tissue-derived stromal cells and a biodegradable scaffold, *Journal of biomedical materials research. Part B, Applied biomaterials 76*, 230-239.

[13] Kakudo, N., Shimotsuma, A., Miyake, S., Kushida, S., and Kusumoto, K. (2008) Bone tissue engineering using human adipose-derived stem cells and honeycomb collagen scaffold, *Journal of biomedical materials research. Part A 84*, 191-197.
[14] Wei, Y., Hu, Y., Hao, W., Han, Y., Meng, G., Zhang, D., Wu, Z., and Wang, H. (2008) A novel injectable scaffold for cartilage tissue engineering using adipose-derived adult stem cells, *Journal of orthopaedic research : official publication of the Orthopaedic Research Society 26*, 27-33.
[15] Awad, H. A., Wickham, M. Q., Leddy, H. A., Gimble, J. M., and Guilak, F. (2004) Chondrogenic differentiation of adipose-derived adult stem cells in agarose, alginate, and gelatin scaffolds, *Biomaterials 25*, 3211-3222.
[16] Betre, H., Ong, S. R., Guilak, F., Chilkoti, A., Fermor, B., and Setton, L. A. (2006) Chondrocytic differentiation of human adipose-derived adult stem cells in elastin-like polypeptide, *Biomaterials 27*, 91-99.
[17] Dragoo, J. L., Carlson, G., McCormick, F., Khan-Farooqi, H., Zhu, M., Zuk, P. A., and Benhaim, P. (2007) Healing full-thickness cartilage defects using adipose-derived stem cells, *Tissue engineering 13*, 1615-1621.
[18] Jin, X. B., Sun, Y. S., Zhang, K., Wang, J., Shi, T. P., Ju, X. D., and Lou, S. Q. (2008) Tissue engineered cartilage from hTGF beta2 transduced human adipose derived stem cells seeded in PLGA/alginate compound in vitro and in vivo, *Journal of biomedical materials research. Part A 86*, 1077-1087.
[19] Jung, Y., Chung, Y. I., Kim, S. H., Tae, G., Kim, Y. H., and Rhie, J. W. (2009) In situ chondrogenic differentiation of human adipose tissue-derived stem cells in a TGF-beta1 loaded fibrin-poly(lactide-caprolactone) nanoparticulate complex, *Biomaterials 30*, 4657-4664.
[20] Mehlhorn, A. T., Zwingmann, J., Finkenzeller, G., Niemeyer, P., Dauner, M., Stark, B., Sudkamp, N. P., and Schmal, H. (2009) Chondrogenesis of adipose-derived adult stem cells in a poly-lactide-co-glycolide scaffold, *Tissue engineering. Part A 15*, 1159-1167.
[21] Jin, X., Sun, Y., Zhang, K., Wang, J., Shi, T., Ju, X., and Lou, S. (2007) Ectopic neocartilage formation from predifferentiated human adipose derived stem cells induced by adenoviral-mediated transfer of hTGF beta2, *Biomaterials 28*, 2994-3003.
[22] Chang, J. C., Hsu, S. H., and Chen, D. C. (2009) The promotion of chondrogenesis in adipose-derived adult stem cells by an RGD-chimeric protein in 3D alginate culture, *Biomaterials 30*, 6265-6275.
[23] Cheng, N. C., Estes, B. T., Awad, H. A., and Guilak, F. (2009) Chondrogenic differentiation of adipose-derived adult stem cells by a porous scaffold derived from native articular cartilage extracellular matrix, *Tissue engineering. Part A 15*, 231-241.
[24] Hildner, F., Concaro, S., Peterbauer, A., Wolbank, S., Danzer, M., Lindahl, A., Gatenholm, P., Redl, H., and van Griensven, M. (2009) Human adipose-derived stem cells contribute to chondrogenesis in coculture with human articular chondrocytes, *Tissue engineering. Part A 15*, 3961-3969.
[25] Patrick, C. W., Jr., Chauvin, P. B., Hobley, J., and Reece, G. P. (1999) Preadipocyte seeded PLGA scaffolds for adipose tissue engineering, *Tissue engineering 5*, 139-151.
[26] Lee, J. A., Parrett, B. M., Conejero, J. A., Laser, J., Chen, J., Kogon, A. J., Nanda, D., Grant, R. T., and Breitbart, A. S. (2003) Biological alchemy: engineering bone and fat from fat-derived stem cells, *Annals of plastic surgery 50*, 610-617.

[27] von Heimburg, D., Zachariah, S., Low, A., and Pallua, N. (2001) Influence of different biodegradable carriers on the in vivo behavior of human adipose precursor cells, *Plastic and reconstructive surgery 108*, 411-420; discussion 421-412.

[28] Huss, F. R., and Kratz, G. (2001) Mammary epithelial cell and adipocyte co-culture in a 3-D matrix: the first step towards tissue-engineered human breast tissue, *Cells, tissues, organs 169*, 361-367.

[29] Hong, L., Peptan, I. A., Colpan, A., and Daw, J. L. (2006) Adipose tissue engineering by human adipose-derived stromal cells, *Cells, tissues, organs 183*, 133-140.

[30] Halbleib, M., Skurk, T., de Luca, C., von Heimburg, D., and Hauner, H. (2003) Tissue engineering of white adipose tissue using hyaluronic acid-based scaffolds. I: in vitro differentiation of human adipocyte precursor cells on scaffolds, *Biomaterials 24*, 3125-3132.

[31] Mizuno, H., Itoi, Y., Kawahara, S., Ogawa, R., Akaishi, S., and Hyakusoku, H. (2008) In vivo adipose tissue regeneration by adipose-derived stromal cells isolated from GFP transgenic mice, *Cells, tissues, organs 187*, 177-185.

[32] Vallee, M., Cote, J. F., and Fradette, J. (2009) Adipose-tissue engineering: taking advantage of the properties of human adipose-derived stem/stromal cells, *Pathologie-biologie 57*, 309-317.

[33] Yoshimura, K., Sato, K., Aoi, N., Kurita, M., Hirohi, T., and Harii, K. (2008) Cell-assisted lipotransfer for cosmetic breast augmentation: supportive use of adipose-derived stem/stromal cells, *Aesthetic plastic surgery 32*, 48-55; discussion 56-47.

[34] Yoshimura, K., Sato, K., Aoi, N., Kurita, M., Inoue, K., Suga, H., Eto, H., Kato, H., Hirohi, T., and Harii, K. (2008) Cell-assisted lipotransfer for facial lipoatrophy: efficacy of clinical use of adipose-derived stem cells, *Dermatologic surgery : official publication for American Society for Dermatologic Surgery [et al.] 34*, 1178-1185.

[35] Stillaert, F. B., Di Bartolo, C., Hunt, J. A., Rhodes, N. P., Tognana, E., Monstrey, S., and Blondeel, P. N. (2008) Human clinical experience with adipose precursor cells seeded on hyaluronic acid-based spongy scaffolds, *Biomaterials 29*, 3953-3959.

[36] Miranville, A., Heeschen, C., Sengenes, C., Curat, C. A., Busse, R., and Bouloumie, A. (2004) Improvement of postnatal neovascularization by human adipose tissue-derived stem cells, *Circulation 110*, 349-355.

[37] Kang, S. K., Shin, M. J., Jung, J. S., Kim, Y. G., and Kim, C. H. (2006) Autologous adipose tissue-derived stromal cells for treatment of spinal cord injury, *Stem cells and development 15*, 583-594.

[38] Banas, A., Teratani, T., Yamamoto, Y., Tokuhara, M., Takeshita, F., Osaki, M., Kawamata, M., Kato, T., Okochi, H., and Ochiya, T. (2008) IFATS collection: in vivo therapeutic potential of human adipose tissue mesenchymal stem cells after transplantation into mice with liver injury, *Stem Cells 26*, 2705-2712.

In: Tissue Engineering
Editors: T. Ishikawa and M. Yamamoto
ISBN 978-1-62257-062-1

Chapter 6

TISSUE ENGINEERING AND REGENERATIVE MEDICINE FOR TREATMENT OF CONGENITAL BIRTH DEFECTS

Hiroaki Komuro
Department of Pediatric Surgery,
Graduate School of Comprehensive Human Sciences,
University of Tsukuba, Tsukuba,
Ibaraki, Japan

ABSTRACT

Tissue engineering and regenerative medicine (TERM) uses a combination of cells, scaffolding and bioreactive factors to treat a variety of pathologic conditions and has become a treatment option for many adult diseases. The use of TERM to treat congenital birth defects has not been well established. However, for the following reasons, it has the potential to be even more successful in pediatric applications than with adults. First, TERM is minimally invasive and a gentle treatment strategy for small infants with congenital birth defects. Second, tissue regeneration in infants is much more vigorous than that in adults because the number of stem and progenitor cells as well as the proliferative capacity of the latter is greater in infants. Third, regenerated tissues are expected to grow with age. This is an important consideration in the treatment of infants. Furthermore, there is no need to take immunosuppressive drugs during growth and maturation of the child. Thus, the use of TERM in infants should be particularly advantageous. The authors investigated the possibility of using TERM in treatments for congenital soft tissue defects that currently require surgical repair, such as esophageal atresia, some bladder anomalies that require augmentation, diaphragmatic hernia, and imperforate anus. Our treatment strategies using TERM for these defects and the results obtained with animal models are presented. These include *in situ* regeneration of esophagus using gastric acellular matrix, *ex vivo* construction of the urinary bladder wall, diaphragmatic repair using a combination of scaffolds and marrow-derived mesenchymal stem cells, and a novel therapy for fecal incontinence after surgical repair of imperforate anus by myoblast transplantation.

Finally, we introduce our study on the potential use of stem cells derived from the amnion as a novel stem cell source for congenital and perinatal diseases. The current status and future prospects of TERM in the treatment of congenital birth defects are reviewed and discussed.

Introduction

Congenital birth defects are commonly encountered. Approximately 3% of newborns have a major congenital anomaly that will affect their quality of life. Congenital birth defects that require surgical reconstruction occur in various organs. When the defect is too large for reconstruction, it is necessary to replace the defect and three basic strategies have been used. The first is to replace the defective organ with an autologous organ. For example, gastrointestinal organs have been used to replace the esophagus in long gap esophageal atresia. They are also used in urinary bladder augmentation. The second strategy is to use artificial organs, including artificial kidneys, vessels, lungs, etc. The third choice is to transplant allogeneic organs. Transplantation of some organs including liver, heart, small intestine, and kidney has been widely performed. However, all of these treatment approaches have significant problems. Replacement by autologous organs often requires highly invasive techniques, and they do not function as well as the original organs, causing many complications. Artificial organs are basically foreign bodies, and there are risks of infection and rejection. It may be necessary to replace artificial organs because of functional failure or inability to grow with the child. Allogeneic transplantation has many disadvantages, including a shortage of donors, rejection, and a requirement for immunosuppressive therapy. Recently, TERM has offered promising strategies for replacement of defective organs, particularly for pediatric patients. First of all, TERM is minimally invasive for small children and the growth potential of manipulated cells should be greater in children than in adults. Furthermore, regenerated organs should grow as the child matures. Accordingly, TERM is a promising strategy for organ replacement in pediatric patients with congenital defects. Below, we discuss recent developments in regenerative medicine.

I. Esophageal Regeneration for the Treatment of Long Gap Esophageal Atresia

Esophageal atresia (EA) consists of a variety of congenital anatomic defects that are caused by abnormal embryological development of the esophagus. EA is anatomically characterized by a congenital obstruction of the esophagus with interrupted continuity of the esophageal wall. The esophagus is divided into two blind pouches, an upper and lower, which may or may not communicate with the tracheobronchial tree through a tracheoesophageal fistula. EA requires surgical repair to reconnect the two ends with or without closure of the fistula. In cases with a short gap, end-to-end esophageal anastomosis is achieved. However, esophageal substitution is required in long gap esophageal atresia, in which the gap between upper and lower esophageal segments might be too long to bridge by direct anastomosis. Although gastrointestinal segments including stomach, small intestine and colon are commonly used in the esophageal substitution, these operations consisting of laparotomy for resection of gastrointestinal segments and esophageal reconstruction are highly invasive. These segments do not function as well as the esophagus as an efficient conduit. Surgery is often followed by postoperative complications including dismotility, anastomotic leakage, stricture, dysphasia, gastroesophageal reflux and future development of cancer, and may lower the quality of life [1-3]. Recently, replacement of the esophageal defect by a tissue engineered esophagus has been considered a promising strategy. These procedures can be performed by less invasive surgery. Many attempts have been applied to esophageal regeneration using *in vivo* as well as *in vitro* tissue engineering [4-9]. The choice of scaffolding is important. Naturally derived extracellular matrices such as porcine-derived small intestinal submucosa (SIS) and bladder acellular matrix (BAM), naturally derived polymers such as collagen, and synthetic polymers including poly-lactic-co-glycolic acid (PLGA) have been used for tissue engineering of the esophagus [4-9]. Naturally derived acellular matrices such as SIS and BAM promote cell migration, cell adherence, and tissue organization *in vivo* [6]. These characteristics would be advantageous in *in situ* tissue engineering, during which repopulation by functional cells is essential. Recently, esophageal acellular matrix (EAM) has been found useful in esophageal tissue engineering [7, 8]. In esophageal regeneration for esophageal atresia, it is difficult to use either BAM or EAM for autografting considering the damage to the urinary bladder and the shortage of esophageal tissue. SIS is a commercially available porcine-derived material, but it is xenogeneic. Accordingly, the potential risk for unknown infections has yet to be solved. Therefore, we proposed the use of a gastric acellular matrix (GAM) for *in situ* tissue engineering of the esophagus. GAM is a naturally derived scaffold which should promote tissue regeneration as do SIS and BAM. GAM is prepared in the same fashion as BAM and should maintain the basement membrane even after removal of cells, which preserves the characteristics of selective permeability [10, 11]. GAM can be prepared as an autograft in a clinical setting using low invasive techniques, since sacrificing a portion of the stomach for GAM can be done without complications. Implanting an autologous scaffold is ideal, because there are no problems with rejection, unknown infections or ethics. GAM is devoid of acid-producing gastric mucosal glands. GAM has been used for *in situ* tissue engineering of urinary bladder and small intestine, and has shown promising results [10,12] When the esophagus is regenerated using GAM, it should grow with the child. Thus, GAM is thought to be an ideal scaffold for esophageal regeneration. Our strategy for esophageal regeneration for EA is shown in Figure 1.

Regeneration of the esophagus using GAM was examined in an inbred rat model. Female to female transplantation of syngeneic grafts (isografts) was used in this study because the lack of rejection should mirror that seen in autografts. GAM patch grafts were implanted in the abdominal esophagus of rats (Figure 2A and 2B). In GAM patch grafts, esophageal mucosal regeneration was excellent (Figure 2C). Early prominent cell proliferation at the bottom of the neomucosa was confirmed by BrdU assay [9]. As early as two weeks after implantation, regeneration of keratinized stratified squamous mucosa was complete. Macroscopically, stenosis or dilatation was not observed in any rat [9]. In contrast, there was no ingrowth of the inner or outer muscle layer detected by desmin at the graft site even after 18 months [9]. Immunohistochemical study showed that the lamina muscularis mucosae did not regenerate at the graft site. Several reports on animal experiments involving *in situ* tissue engineering of the esophagus have shown regeneration or ingrowth of the muscle layer or muscle cells [4-6].

Although the presence of muscle-like cells or myofibroblasts positive for muscle-specific antigen has been shown, a large volume of regenerated muscle tissue has never been observed. In our study, a BrdU-based proliferation assay showed that there was no increased incorporation of BrdU at the margins of the native muscle layer during esophageal regeneration using GAM [9].

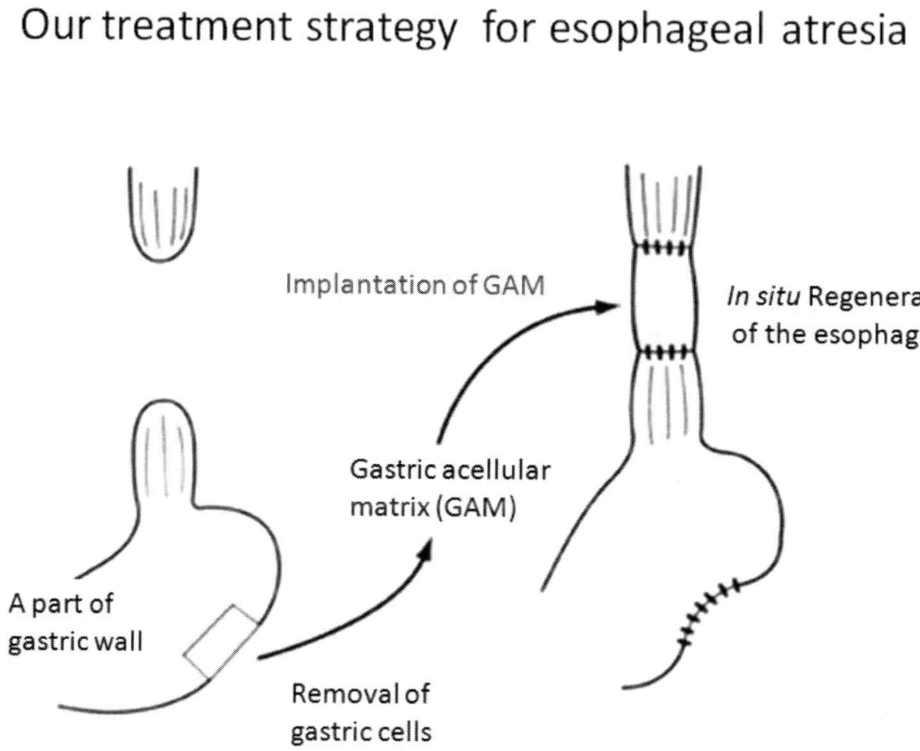

Figure 1. Our treatment strategy for long gap esophageal atresia, using autologous gastric acellular matrix.

Regeneration of the esophagus using GAM was examined in an inbred rat model. Female to female transplantation of syngeneic grafts (isografts) was used in this study because the lack of rejection should mirror that seen in autografts. GAM patch grafts were implanted in the abdominal esophagus of rats (Figure 2A and 2B). In GAM patch grafts, esophageal mucosal regeneration was excellent (Figure 2C). Early prominent cell proliferation at the bottom of the neomucosa was confirmed by BrdU assay [9]. As early as two weeks after implantation, regeneration of keratinized stratified squamous mucosa was complete. Macroscopically, stenosis or dilatation was not observed in any rat [9]. In contrast, there was no ingrowth of the inner or outer muscle layer detected by desmin at the graft site even after 18 months [9]. Immunohistochemical study showed that the lamina muscularis mucosae did not regenerate at the graft site. Several reports on animal experiments involving *in situ* tissue engineering of the esophagus have shown regeneration or ingrowth of the muscle layer or muscle cells [4-6].

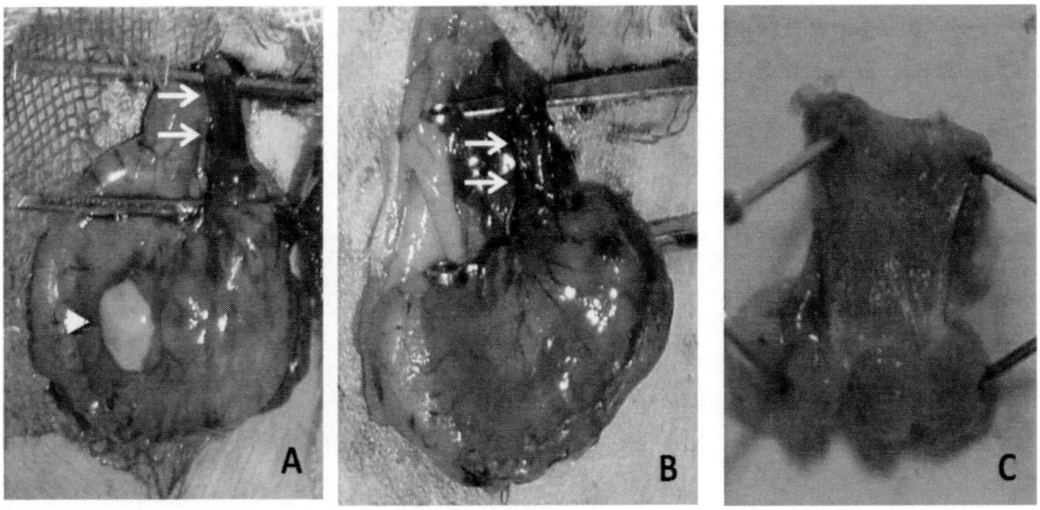

Figure 2. Surgical procedure for the replacement of a rat abdominal esophageal defect and macroscopic findings after mucosal regeneration (A) The defect in the abdominal esophagus is indicated by arrows. Gastric acellular matrix (GAM) patch is shown by an arrow head. (B) Implanted GAM patch graft is shown by arrows. (C) The graft site was completely covered by neomucosa four months after implantation.

The number of BrdU-positive cells was the same as that in the intact muscle. This suggests that muscle stem cells (satellite cells) did not enter into active proliferation and that muscle regeneration did not occur. Further efforts are required to regenerate muscle tissue and the use of growth factors or muscle stem cells is under examination. In these investigations, there is a fundamental problem in the use of animal models. In most animals, muscle layers in the esophagus are composed of skeletal muscles, while in human, they consist of smooth muscles. Thus, esophageal regeneration in animals may be quite different from that in humans. At present, it is not clear whether the muscle layer is essential for the esophagus to function. A small amuscular portion of the esophagus would probably not influence esophageal function as long as the area shows some mechanical strength. Circular myotomy has been conducted in patients with esophageal atresia to elongate the esophagus before anastomosis (Figure 3). Patients with an amuscular portion have generally not shown problems postoperatively, although others have been complicated by ballooning [13]. Therefore, esophageal regeneration without muscle layers using GAM may provide a useful tool for bridging the proximal and distal ends in esophageal atresia showing a long gap.

In summary, GAM appears to be a valuable scaffolding for tissue engineering of the esophagus. The advantage of GAM over other scaffolds is that it can be prepared less invasively as a naturally derived autograft for clinical application. GAM patch grafts provided satisfactory results for esophageal mucosal regeneration without stenosis or dilatation. Our data also suggest that regeneration of the muscle layer remains a challenge. A new breakthrough for muscle regeneration accompanied by nerve regrowth and neovascularization including application of stem cells and growth factors may be required.

II. Urinary Bladder Regeneration for Bladder Augmentation in Congenital Diseases

Urinary bladder augmentation is required in patients with some congenital defects such as spina bifida, bladder extrophy, cloacal extrophy, and microcystis other than surgical resection of bladder malignancy. Gastrointestinal segments are commonly used for augmentation of the urinary bladder. However, the use of gastrointestinal segments for bladder augmentation is associated with multiple complications such as infection, metabolic disorders, calculi formation, increased mucous production and malignancy. In patients who have short gut syndrome or severe adhesions caused by previous operations, gastrointestinal segments may be difficult to use. Gastrointestinal segments should not be sacrificed, particularly in growing children. Tissue-engineering technology with autologous native cells has been applied to urinary bladder augmentation. Reconstruction of the native urinary bladder using autologous cells and biodegradable scaffolding appears to be a promising method for augmentation [14-16].

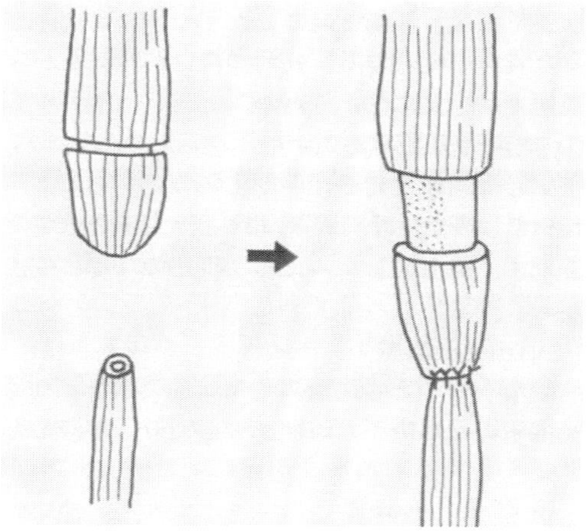

Figure 3. Esophageal elongation by circular myotomy and esophago-esophageal anastomosis.

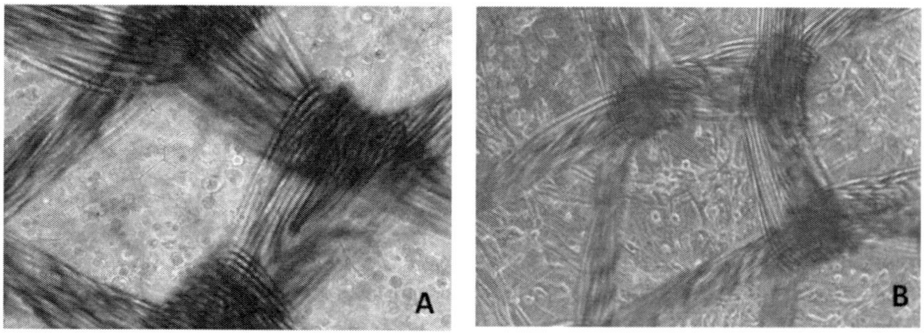

Figure 4. Phase contrast microscopic appearance of the proliferating cells on the hybrid scaffolds. (A) Urothelial cells on the hybrid PLGA mesh-collagen sponge scaffold. (B) Smooth muscle cells on the hybrid PLGA mesh-collagen gel scaffold.

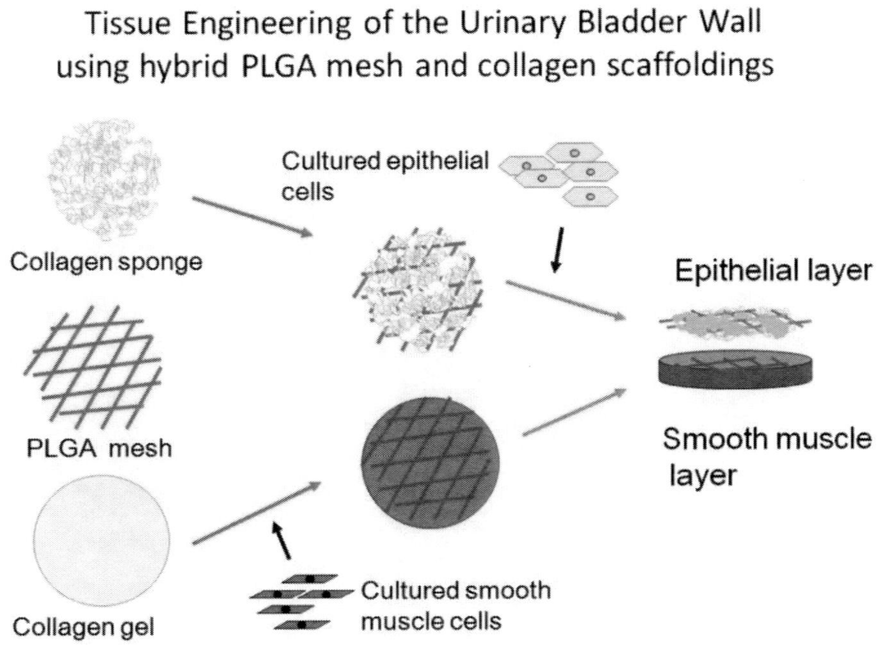

Figure 5. Construction of the tissue engineered urinary bladder wall.

Three-dimensional biodegradable scaffolds play an important role in tissue engineering as an adhesive substrate for implanted cells and a physical support to guide the formation of new organs. The ideal scaffold should have good biocompatibility and mechanical strength for easy handling, and should be good for cell seeding. Generally, synthetic polymers have the proper mechanical strength, but they are relatively hydrophobic and hinder cell seeding. In contrast, naturally derived collagen offers the advantages of hydrophilic cell adhesion and specific cell-to-cell interactions, but scaffolds constructed entirely of collagen have poor mechanical strength. A logical alternative is combining synthetic biodegradable polymers with collagen. Chen *et al.* showed the usefulness of hybrid scaffolding by combining synthetic polymer for mechanical strength with naturally derived collagen to enhance cell seeding [17, 18]. They prepared hybrid scaffoldings from PLGA mesh-collagen sponge and PLGA mesh-collagen gel. These hybrid scaffolds achieved good biocompatibility, mechanical strength, and cell seeding properties, and we used them to reconstruct a tissue-engineered urinary bladder wall. The hybrid PLGA mesh-collagen scaffolds were prepared by introducing collagen sponge or gel into the openings of a PLGA knitted mesh. Urothelial and smooth muscle cells were obtained from porcine urinary bladder wall, and were cultured in their respective media. The urothelial layer was regenerated by culturing urothelial cells in the three dimensional PLGA mesh and collagen sponge (Figure 4A). Scanning electron microscopy revealed good cell-cell contact of the urothelial cells on the scaffold, resembling cobble stones. Microvilli were present on the surface of the stratified urothelial layer, which is specific for apical cells of urothelium [19, 20]. Thus, the urothelial cells cultured on this scaffolding preserved the characteristics of the native urothelial cells. The smooth muscle layer was reconstructed by culturing smooth muscle cells in the PLGA mesh and collagen gel (Figure 4B). A novel tissue-engineered urinary bladder wall was constructed by laminating the urothelial and smooth muscle layers (Figure 5) [19, 20].

When we implanted the tissue-engineered urinary bladder wall to the back skin of nude mice, engraftment was not achieved. Atala *et al.* succeeded in creating a tissue-engineered urinary bladder using cells obtained from patients and a collagen-PLGA composite scaffold [18]. They performed omental wrapping to supply vascularization when implanted. However, no other successful results using similar strategies have been reported and obstacles remain. It is difficult to distinguish whether the regenerated bladder functioned from the possibility that the residual native bladder compensated for the lost function. We suggest that novel techniques should be developed to more efficiently promote the blood supply. Furthermore, to achieve bladder function (low-pressure storage and emptying by contraction), re-innervation by neural generation is essential. It is not clear whether the previously reported tissue engineered bladders were well innervated. Further efforts are needed because there are great expectations for restoring urinary bladder function through regenerative medicine.

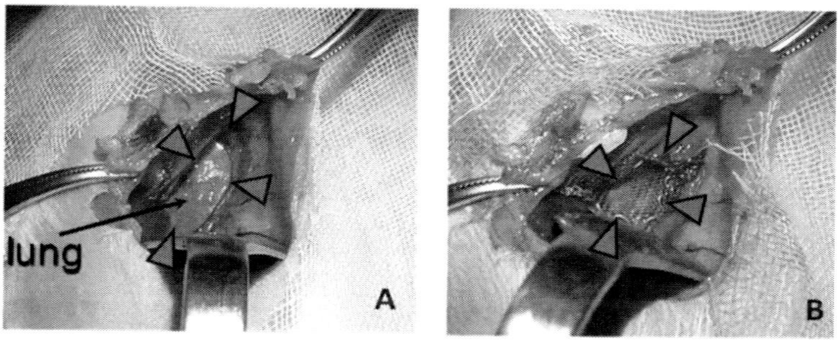

Figure 6. Surgical procedure of diaphragmatic repair using a patch graft in rats. (A) A defect was created in the left diaphragm (arrow heads). The lung is shown by an arrow. (B) The defect was repaired using a patch graft (shown by arrow heads).

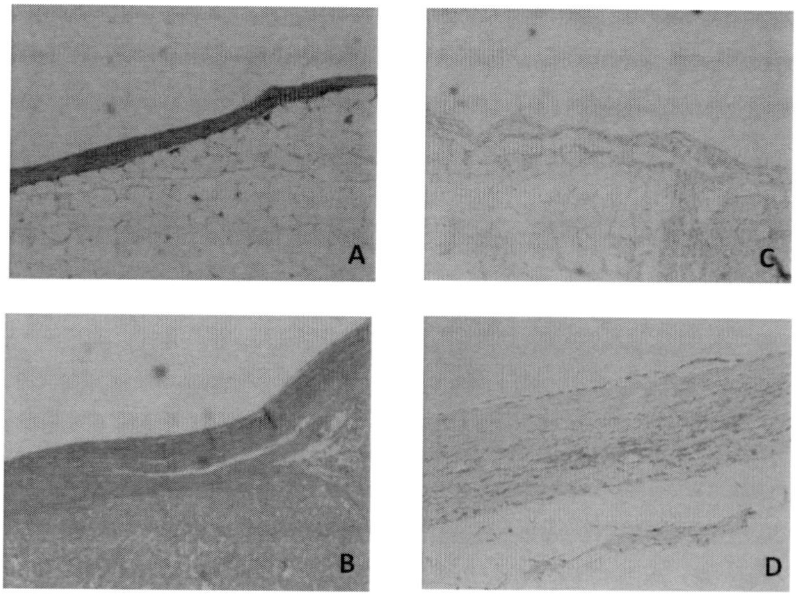

Figure 7. Microscopic findings of the patch grafts. The graft sites were replaced by a thin fibrous tissue in group 1 (A) and by a much thicker one in group 2 (B) (Masson's Trichrome stain). Immunohistochemical staining shows the scattered desmin-positive cells in group 2 (D), but not in group 1 (C).

Table 1. Results of rat diaphragmatic repair using scaffolds

	Type of scaffold	Thickness of neo-diaphragm	Presence of desmin-positive cells
Group 1	PLGA mesh	thin	(-)
Group 2	PLGA mesh+collagen	thick	(+)
Group 3	PLGA mesh+collagen+MSC	thick	(+)

MSC, mesenchymal stem cells.

III. Regeneration of the Diaphragm for Congenital Diaphragmatic Hernia Repair

Congenital diaphragmatic hernia (CDH) is a congenital defect in the diaphragm that allows most of the abdominal contents to herniate into the chest cavity. CDH is a life-threatening disorder because the lungs are hypoplastic and pulmonary hypertension develops. The outcome of surgery depends on the degrees of lung hypoplasia. High mortality rates have been reported especially in prenatally diagnosed cases and the severest cases may require fetal surgery [21-24]. Diaphragmatic hernia repair may be done by direct closure, but larger defects may require patch closure. In severe CDH, the defect can be quite large. Prosthetic patches have been routinely used to close large diaphragmatic defects in neonates and infants. Implantation of these synthetic materials exposes patients to the risks of rejection, granulation, infection, allergic reaction and recurrence due to patch displacement or repair disruption and also chest wall deformity in long-time survivors. On the other hand, recent advances in the perinatal care of infants with severe CDH have dramatically lowered their mortality rates. Consequently, the number of survivors with severe CDH who require a prosthetic patch for larger defects has been increasing. This in turn has led to an increase in recurrence rates following repair [25-28]. We postulated that in order to lower the risk of morbidity and recurrence, an autologous, tissue-engineered patch might be of benefit. We evaluated the efficacy of diaphragmatic hernia repair in a rat model using a hybrid PLGA mesh-collagen sponge scaffolding, designed for *in situ* tissue engineering [29]. Oval-shaped defects were surgically created in the left diaphragm and repaired with three different grafts: PLGA mesh in group 1, nybrid PLGA mesh-collagen sponge scaffolding in group 2, and hybrid PLGA mesh-collagen sponge scaffolding seeded with syngeneic bone marrow-derived mesenchymal stem cells (MSCs) in group 3 (Figure 6). All patch grafts examined were viable without evidence of necrosis and maintained structural integrity without eventration. After one month, we observed in each group a membranous lining of fibrous tissue over the scaffold, and also noted inflammatory cell infiltration into and around the scaffold. The inflammatory findings disappeared after two months. PLGA mesh was partly absorbed with blurring of the original structure within a month, and the basic structure of the PLGA mesh almost disappeared after two or three months. Ingrowth of fibroblasts into the graft was observed in all groups. The mean thickness around the center of the 'neodiaphragm' was 200 μm in group 1, and 550 μm in groups 2 and 3. The mean marginal thickness of the 'neodiaphragm' was 600 μm in group 1, 1000 μm in group 2, and 1250 μm in group 3. The regenerated tissues in groups 2 and 3 were significantly thicker than that in group 1 in the central as well as marginal portions ($P < 0.05$) (Figure 7A and 7B). Although regeneration of muscular structure could not be observed in any of the three groups even three months after the operation, scattered desmin-positive cells were found within the graft in groups 2 and 3 (Figure 7C and 7D). There was little difference between groups 2 and 3 as far as microscopic findings, indicating no benefit of MSC seeding in the scaffold. The desmin-positive cells were likely of host origin, because they were found without MSC seeding. The experimental results are summarized in Table 1. The seeding of MSC on the hybrid scaffold did not bring about any advantage in the present histological study, although functional evaluations were not performed. Our treatment strategy using the hybrid scaffolds allows CDH repair by autologous tissues without residual foreign bodies for the large defects. These regenerated tissues are expected to grow with the patients. Although further validation of its usefulness in CDH repair in growing animals may be required, our study suggests that hybrid PLGA mesh-collagen sponge scaffolding offers a useful approach to autologous *in situ* tissue regeneration and growth.

IV. Myoblast Transplantation for the Treatment of Fecal Incontinence in Imperforate Anus

Imperforate anus consists of a spectrum of congenital defects in which the anus has not developed in the normal position. Most patients with imperforate anus have abnormal defecation muscles with various degrees of underdevelopment, and will experience some degree of fecal incontinence after surgical repair. Imperforate anus is classified into low, intermediate, and high types. The level of atresia is an important prognostic factor in terms of postoperative fecal continence. The obvious cause of poor fecal continence in high type anomalies is the marked hypoplasia of the voluntary sphincter muscles [30]. There has been no effective treatment for such postoperative fecal incontinence. Now, we speculate that muscle regeneration using muscle-derived progenitor cells may provide a potential treatment strategy to improve fecal incontinence after surgical repair of imperforate anus. Figure 8 shows the process of muscle differentiation. Muscle stem cells differentiate into desmin-positive myoblasts (muscle progenitor cells) which then fuse to become multinucleated myotubes with expression of myosin. Finally, they mature to become contractile muscle fibers with a striated structure. Myoblasts can be prepared from a small muscle biopsy and can be easily amplified *in vitro*. They survive and differentiate into multinucleated muscle fibers when transplanted. The use of autologous myoblasts overcomes problems related to ethics as well as immune rejection. Because they are at an advanced stage of differentiation along the myogenic pathway, they are unlikely to form tumors when transplanted. They also show a high resistance to ischemia, which suggests the efficacy of transplantation in recipients with poor vascular environments. Figure 9 shows our treatment strategy for postoperative fecal incontinence in patients with imperforate anus. Following collection of a small muscle biopsy, myoblasts are isolated, cultured and amplified *in vitro*. Then, they are transplanted into the defecation muscle complex. Finally, they reinforce the underdeveloped muscle complex by muscle regeneration.

We investigated the efficacy of rat myoblast preparations of cultured myoblasts at different timings by assessing the frequency of myoblasts following culture. The efficiency of myoblast preparation was not very high at birth, but it increased after one to two weeks, and thereafter, it decreased with age [31]. This suggested the effectiveness of myoblast preparation in young infants with imperforate anus. Furthermore, it is anticipated that the cells will differentiate efficiently when introduced to young infants. Next, we transplanted cultured myoblasts to the defecation muscles in rats. The levator ani muscle was exposed with a transverse skin incision back from the anus (Figure 10) and GFP-positive myoblasts were injected into GFP-negative rats. The transplanted myoblasts surrounding the levator ani muscle survived and expressed myosin (myotube formation). However, myofiber formation by transplanted myoblasts was not observed thereafter [31]. Some further stimuli might be required to form myofibers. In clinical situations, the defecation muscles will have some mechanical damage from the various surgical procedures carried out for repair of the imperforate anus which may stimulate the transplanted myoblasts to differentiate when introduced at surgery. In previous experiments examining the transplantation of myoblasts into the urethral sphincter, various types of sphincter injury models, such as denervation, ischemia, cauterizing, resection, injection of toxin, and freezing, have been used [32-36]. Urinary incontinence results from sphincter deficiency due to aging or an acquired injury such as obstetrical trauma and surgical damage. In our study, some damage to the levator ani muscle might have been required to induce myofiber formation, although ectopic muscle formation derived from myoblasts in the skin has been reported [37]. The addition of growth factors, such as basic fibroblast growth factor or nerve growth factor, might also stimulate myoblasts [38]. We are currently planning experiments using a rat model in which slight surgical stress or biological stimuli are used in combination with transplanted myoblasts.

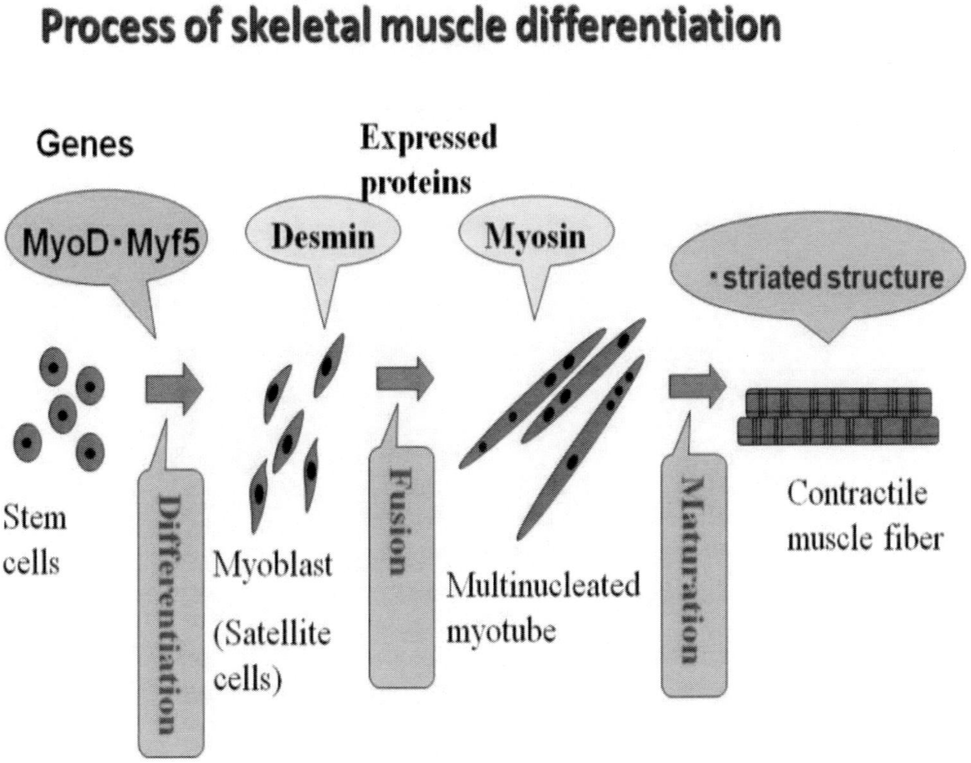

Figure 8. Process of skeletal muscle differentiation.

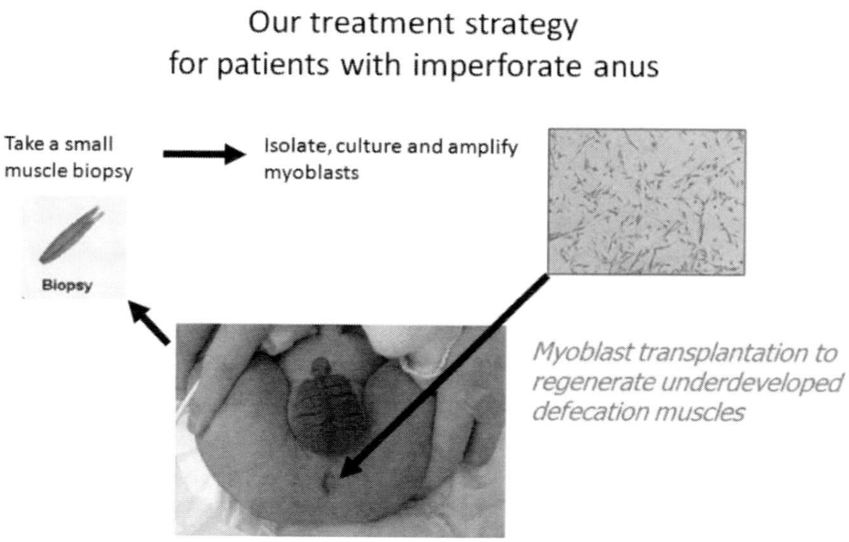

Figure 9. Our novel treatment strategy for fecal incontinence after the surgical repair of imperforate anus using myoblast transplantation.

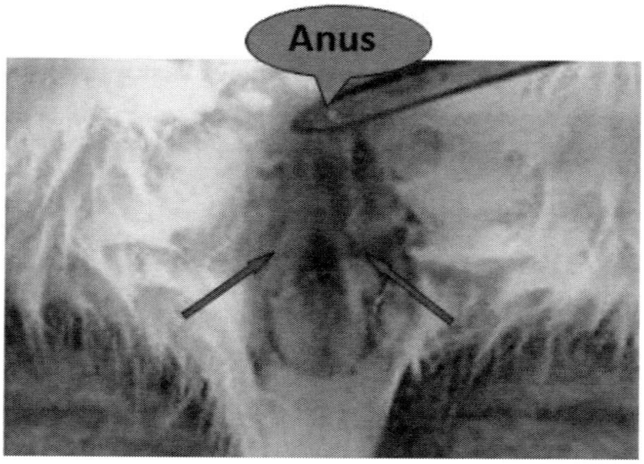

Figure 10. Procedure for myoblast transplantation into the levator ani muscle in rats.

Figure 11. Pig models of imperforate anus.

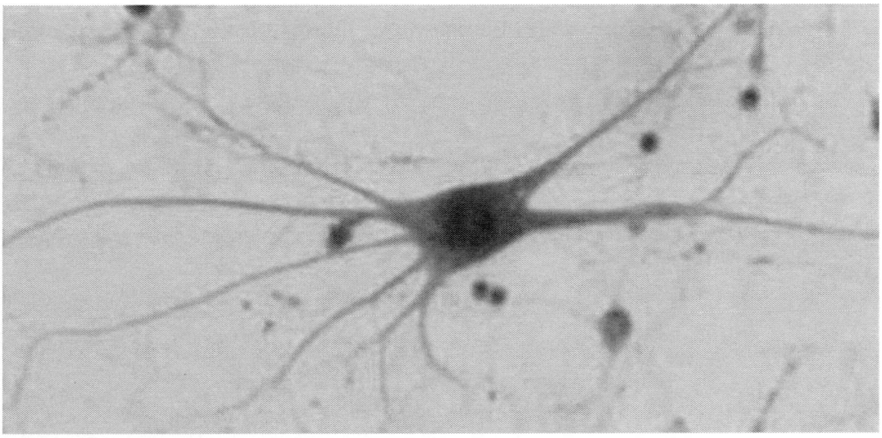

Figure 12. βIII-tublin expression in rat amniotic epithelial cells after neuronal induction. The cells developed neuron-like extensions.

Colleagues have recently generated a porcine line in which imperforate anus occurs naturally (Figure 11) [39]. We are planning to perform preclinical experiments using these pig models in the near future.

In summary, myoblasts are readily amplified and can be successfully introduced into the levator ani muscle in rats, resulting in formation of a bulk of myosin-expressing myotubes. These findings strongly suggest the potential effectiveness of autologous myoblast transplantation in young infants with imperforate anus. Although further efforts are required to produce myofiber formation, autologous myoblast transfer therapy using a small muscle biopsy may offer a novel approach for the treatment of fecal incontinence after repair of imperforate anus.

V. Amniotic Tissue as a Potential Source of Stem Cells

The utilization of stem cells is an important aspect of regenerative medicine. Although induced pluripotent stem (iPS) cells have been recently developed and are in the spotlight, there remain many problems to be resolved for their clinical application, including reducing the chance of cancer development [40, 41]. On the other hand, naturally derived stem cells are generally classified into two major subtypes: embryonic stem (ES) cells (derived from the inner cell mass of the preimplantation embryo), and somatic stem cells (derived from organ-specific adult stem cells). ES cells are known to be the most versatile pluripotent stem cells, with the ability to *differentiate* into all derivatives of the three primary *germ layer* (*ectoderm*, *endoderm*, and *mesoderm*). They are relatively easy to cultivate in large numbers. Because of their pluripotency and unlimited capacity for self-renewal, great attention has been paid to ES cell therapies. However, their preparation from fertilized eggs carries ethical problems. ES cells are basically allogeneic, and many problems, including rejection, purification, malignant changes, and methods for culturing without animal components, remain to be resolved to enable their clinical application. Somatic stem cells have been shown to have a limited but wide spectrum of differentiation potential across cell lineage boundaries. Unlike ES cells, their use generally avoids ethical problems, and their autologous transplantation is a promising option. However, somatic stem cells constitute only a small fraction of the cells in a tissue and their efficient use is not easy. It remains to be seen whether somatic stem cells are as promising as ES cells.

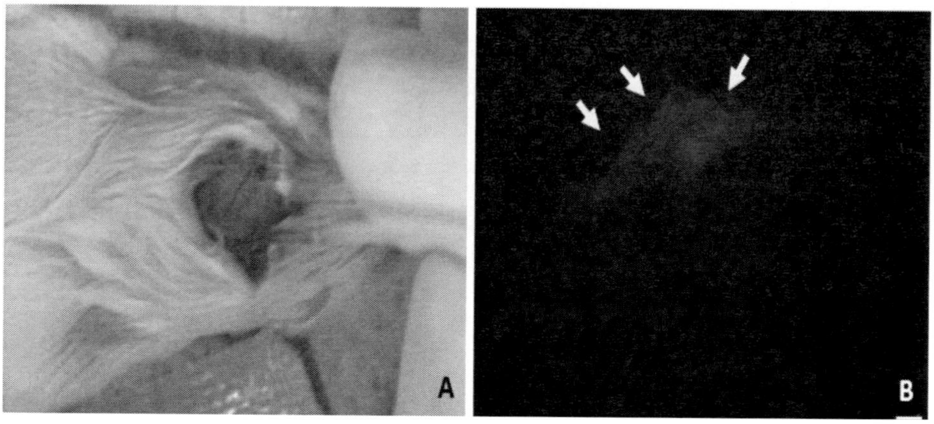

Figure 13. Transplantation of rat amniotic epithelial cells (AEC) into the ischemic brain. (A) Rat AEC labeled by PKH were introduced into the rat ischemic brain. (B) Introduced cells (red) survived two weeks after transplantation. The nuclei in the brain are shown by DAPI (blue).

The amnion is a part of the placenta which is usually discarded after delivery. However, the amnion is derived from the epiblast of the inner cell mass which can generate ES cells. Because of their ability to differentiate, amniotic epithelial cells (AEC) have also been hypothesized to retain pluripotency. Indeed, recent studies have shown that human AEC have the potential to differentiate into endoderm (liver, pancreas), mesoderm (cardiomyocytes) and ectoderm (neural cells) [42-45]. The use of AEC for stem cell therapy has many advantages over the use of other stem cells.First, many AEC can be prepared non-invasively after delivery. Their preparation can avoid ethical restrictions, because they are otherwise discarded. Unlike allogeneic ES cell transplantation, autologous AEC transplantation can be applied to neonatal disorders. Furthermore, AEC can be used for allogeneic transplantation, because of their low immunogenicity. AEC show low expression of major histocompatibility complex (MHC) class I antigens and a lack of MHC class II antigens [46-48]. Their application to fetal cell therapy or fetal tissue engineering might be possible if they could be obtained by amniocentesis and expanded. Thus, the amnion might be a promising alternative source of stem cells. We should mention drawbacks to the use of AEC. In the case of autologous transplants, if genetic defects are the cause of pathology, AEC might not be useful. In the case of allogeneic transplants, there are issues of communicable disease and the expense of banking such cells.

The amnion develops from the epiblast, the anteromedial part of which is further specified to form the neural plate. Many previous studies have shown that AEC are likely to differentiate into neural lineages [42, 45, 49-54]. AEC transfer therapy might then represent an attractive new treatment strategy for perinatal neurogenic disorders. We investigated the neural-differentiation potential of rat amniotic epithelial cells by performing *in vitro* differentiation experiments. In RT-PCR analysis, rat AEC were found to express vimentin and nestin, but not c-kit, oct-4 or nanog. In cultured rat AEC, RT-PCR detected the expression of neural lineage markers, including βIII-tubulin, neuron specific enolase (NSE), neurofilament-M, neuroD, glial fibrillary acidic protein (GFAP), myelin basic protein (MBP), tyrosine hydroxylase (TH), acetylcholinesterase (AChE), choline acetyltransferase (ChAT), and mammalian achaete-scute homolog1 (MASH1) [55]. When rat AEC cells were subjected to neural induction by the method of Woodbury *et al.* [56], they dramatically changed their shapes, projecting dendrite-like structures. The induced cells expressed an immature neuronal marker, βIII-tubulin (Figure 12). Marker proteins specific for mature neurons, astrocytes or oligodendrocytes were not detected using this protocol. Our findings suggested that rat AEC might be already committed to differentiate to various neural lineages and that they could differentiate to immature neurons *in vitro*. Sakuragawa *et al.* showed that human AEC expressed markers of neuronal and glial cells [42, 46]. By further improving protocols, it might be possible to improve neuronal differentiation of AEC.

In the case of perinatal brain injury, the affected neonates have a high mortality rate, while survivors suffer long-term sequelae such as cerebral palsy, seizures, mental retardation, and learning disabilities. Preterm infants are at higher risk of brain injuries including ischemia, intraventricular hemorrhage (IVH), periventricular leukomalacia (PVL) and ventriculomegaly. Despite recent advances in neonatal care, the incidence of perinatal brain injury is not decreasing partly because the number of survivors from preterm delivery is increasing. No effective therapy or way of prevention has yet been developed. Recently, stem cell therapies for neurogenic diseases, including perinatal brain injury have been a focus of attention [57] and it is possible that ES cells could be used as neurogenic stem cells. However, such cells are allogeneic to the patient, and ethical limitations must be resolved. Alternatively, it might be possible to use neural stem cells. They reside in the embryo, and in neurogenic regions of the adult brain: the subventricular zone of the lateral ventricle and the dentate gyrus subgranular zone of the hippocampus. However, their autotransplantation is impossible because their preparation from the brain is highly invasive. In contrast, usage of AEC for perinatal neurogenic disorders is attractive as well as practical because abundant AEC could be obtained non-invasively at delivery and be made available soon after delivery. Previous animal studies have shown the possibilities of AEC therapy for ischemic brain and cord injuries as well as cell-mediated gene therapy using AEC for mucopolysaccharidosis type VII [58-60].

We recently initiated a new experimental study in which AEC were transplanted into the ischemic rat brain. A rat neonatal ischemic brain model was created by ligation of one common carotid artery followed by hypoxia (8% O_2) [61]. The labeled cultured rat AECs were introduced directly into the ipsilateral ischemic brain. The transplanted rat AECs survived in the brain two weeks after transplantation (Figure 13). Future experiments will determine whether these cells can differentiate into neural cells.

In summary, although rat AEC may represent a heterogeneous cell population, our study suggested not only that they expressed neural progenitor gene markers such as nestin and vimentin, but also that they showed the potential to differentiate into neuron, glia, and other neuronal lineages, including catecholaminergic, cholinergic and enteric nervous lineages. Actually, it was confirmed that rat AEC differentiated at least into immature neurons after neural induction *in vitro*. In the future, cell therapy with neurogenic AEC could provide a promising treatment strategy for perinatal neurogenic disorders.

Stem cell sources in extra-embryonic fetal tissues

- ***Umbilical cord***
  - Cord blood-derived hematopoietic stem cells
  - Cord blood-derived mesenchymal stem cells
  - Wharton's jelly-derived mesenchymal stem cells
- ***Placenta***
  - Placenta-derived mesenchymal stem cells (Fetal or Maternal)
- ***Amniotic membrane***
  - Stem cells derived from epithelial layer
  - Stem cells derived from mesenchymal layer
- ***Amniotic fluid***
  - Amniotic fluid-derived stem cells (Unknown origin)

Figure 14. Stem cells identified from extra-embryonic fetal tissues.

Future Directions

The extra-embryonic fetal tissues such as umbilical cord, placenta, amniotic membrane and amniotic fluid are attractive as putative stem cell sources, and a number of types of stem cells have been identified in those tissues (Figure 14). Umbilical cord blood contains hematopoietic stem cells (which are in current clinical use) and mesenchymal stem cells [62, 63]. Wharton's jelly, which composes the matrix of the umbilical cord, has been shown to contain multi-lineage stem cells [64, 65]. Placenta also contains mesenchymal stem cells, although they may be of fetal or maternal origin [66, 67]. Amniotic fluid-derived multipotent stem cells have been demonstrated, although their origin is not known [68]. These might be potentially used for fetal TERM. In humans, amniotic membrane contains stem cells derived not only from the epithelial layer but also from the mesenchymal layer [69-71]. These extra-embryonic fetal stem cells might possess greater pluripotency and cell proliferative capacity than somatic stem cells. Furthermore, with the exception of amniotic fluid-derived stem cells, these cells can be prepared non-invasively and abundantly at delivery, although a potential problem is maintenance of sterility following vaginal delivery. Ethical concerns are limited because they are otherwise discarded. These could be used for autologous transplantation during postnatal life, or they could possibly be used for fetal therapies. Clinical application of stem cells derived from these extra-embryonic fetal tissues would provide an attractive and promising strategy to TERM for congenital birth defects.

References

[1] Hamza, A. F. (2009). Colonic replacement *Semin. Pediatr. Surg.*, 18, 40-43.

[2] Hunter, C. J., Petrosyan, M., Connelly, M. E., Ford, H. R. and Nguyen, N. X. (2009). Repair of long-gap esophageal atresia *Pediatr. Surg. Int.*

[3] Deurloo, J. A., Ekkelkamp, S., Hartman, E. E., Sprangers, M. A. and Aronson, D. C. (2005). Quality of life in adult survivors of correction of oesophageal atresia. *Arch. Surg., 140*, 976-980.

[4] Takimoto, Y., Nakamura, T., Yamamoto, Y., Kiyotani, T., Teramachi, M. and Shimizu, Y. (1998). The experimental replacement of a cervical esophageal segment with an artificial prosthesis with the use of collagen matrix and a silicone stent. *J. Thorac. Cardiovasc. Surg.* 116: 98-106.

[5] Lynen Jansen, P., Klinge, U., Anurov, M., Titkova, S., Mertens, P. R. and Jansen, M. (2004). Surgical mesh as a scaffold for tissue regeneration in the esophagus. *Eur. Surg. Res., 36*, 104-111S.

[6] Badylak, S., Meurling, S., Chen, M., Spievack, A. and Simmons-Byrd, A. (2000). Resorbable bioscaffold for esophageal repair in a dog model. *J. Pediatr. Surg.,* ***35,*** 1097-1103.

[7] Bhrany, A. D., Beckstead, B. L., Lang, T. C., Farwell, D. G., Giachelli, C. M. and Ratner, B. D. (2006). Development of an esophagus acellular matrix tissue scaffold. *Tissue Eng.,* ***12,*** 319-330.

[8] Ozeki, M., Narita, Y., Kagami, H., Ohmiya, N., Itoh, A., Hirooka, Y., Niwa, Y., Ueda, M. and Goto, H. (2006). Evaluation of decellularized esophagus as a scaffold for cultured esophageal epithelial cells. *J. Biomed. Mater. Res. A, 79,* 771-778.

[9] Urita, Y., Komuro, H., Chen, G., Shinya, M., Kaneko, S., Kaneko, M. and Ushida, T. (2007). Regeneration of the esophagus using gastric acellular matrix: an experimental study in a rat model. *Pediatr. Surg. Int. 23,* 21-26.

[10] Meezan, E., Hjelle, J. T. and Brendel, K. (1975). A simple., versatile., nondisruptive method for the isolation of morphologically and chemically pure basement membranes from several tissues. *Life Sci.,17,* 1721-1732.

[11] Parnigotto, P. P., Marzaro, M., Artusi, T., Perrino, G. and Conconi, M. T. (2000). Short bowel syndrome: experimental approach to increase intestinal surface in rats by gastric homologous acellular matrix.*J. Pediatr. Surg., 35,* 1304-1308.

[12] Sutherland, R. S., Baskin, L. S., Hayward, S. W. and Cunha, G. R. (1996). Regeneration of bladder urothelium., smooth muscle., blood vessels and nerves into an acellular tissue matrix. *J. Urol., 156,* 571-577.

[13] Livaditis, A., Radberg, L. and Odensjo, G. (1972). Esophageal end-to-end anastomosis. Reduction of anastomotic tension by circular myotomy. *Scand. J. Thorac. Cardiovasc. Surg., 6,* 206-214.

[14] Oberpenning, F., Meng, J., Yoo, J. J. and Atala, A. (1999). De novo reconstitution of a functional mammalian urinary bladder by tissue engineering. *Nat. Biotechnol., 17,* 149-155.

[15] Fauza, D. O., Fishman, S. J., Mehegan, K. and Atala, A. (1998). Videofetoscopically assisted fetal tissue engineering: bladder augmentation. *J. Pediatr. Surg., 33,* 7-12.

[16] Atala, A., Bauer, S. B., Soker, S., Yoo, J. J. and Retik, A. B. (2006). Tissue-engineered autologous bladders for patients needing cystoplasty. *Lancet, 367(9518)*, 1241-6.

[17] Chen, G., Ushida, T. and Tateishi, T. (2000). A hybrid network of synthetic polymer mesh and collagen sponge. *J. Chem. Soc. Chem. Commun., 16*, 1505-1506.

[18] Tateishi, T., Chen, G. and Ushida, T. (2002). Biodegradable porous scaffolds for tissue engineering. *J. Artif. Organs, 5,* 77-83.

[19] Nakanishi, Y., Chen, G., Komuro, H., Ushida, T., Kaneko, S., Tateishi, T. and Kaneko, M. (2003). Tissue-engineered urinary bladder wall using PLGA mesh-collagen hybrid scaffolds: a comparison study of collagen sponge and gel as a scaffold. *J. Pediatr. Surg., 38,* 1781-1784.

[20] Komuro, H., Nakanishi, Y., Chen, G., Kaneko, M., Tateishi, T. and Ushida, T. (2004). Tissue-engineered urinary bladder wall using scaffold construction. *Am. J. Urol. Rev. 2,* 289-292.

[21] Reickert, C. A., Hirschl, R. B., Atkinson, J. B., Dudell, G., Georgeson, K., Glick, P., Greenspan, J., Kays, D., Klein, M., Lally, K. P., Mahaffey, S., Ryckman, F., Sawin, R., Short, B. L., Stolar, C. J., Thompson, A. and Wilson, J. M. (1998). Congenital diaphragmatic hernia *Surgery., 123,* 305-310.

[22] Skari, H., Bjornland, K., Haugen, G., Egeland, T. and Emblem, R. (2000). Congenital diaphragmatic hernia: a meta-analysis of mortality factors. *J. Pediatr. Surg., 35,* 1187-1197.

[23] Stevens, T. P., van Wijngaarden, E., Ackerman, K. G., Lally, P. A. and Lally, K. P. (2009). Congenital Diaphragmatic Hernia Study Group. Timing of delivery and survival *Pediatrics., 123,* 494-502.

[24] de Buys Roessingh, A. S. and Dinh-Xuan, A. T. (2009). Congenital diaphragmatic hernia *Eur. J. Pediatr., 168,* 393-406.

[25] Sydorak, R. M., Hoffman, W., Lee, H., Yingling, C. D., Longaker, M., Chang, J., Smith, B., Harrison, M. R. and Albanese C. T. (2003). Reversed latissimus dorsi muscle flap for repair of recurrent congenital diaphragmatic hernia. *J. Pediatr. Surg.,* ***38,*** 296-300.

[26] Scaife, E. R., Johnson, D. G., Meyers, R. L., Johnson, S. M. and Matlak, M. E. (2003). The split abdominal wall muscle flap—a simple., mesh-free approach to repair large diaphragmatic hernia. *J. Pediatr. Surg.,* ***38,*** 1748-1751.

[27] Rosenkrantz, J. G. and Cotton, E. K. (1964). Replacement of left hemidiaphragm by a pedicled abdominal muscular flap. *J. Thorac. Cardiovasc. Surg.,* ***48,*** 912-920.

[28] Moss, R. L., Chen, C. M. and Harrison, M. R. (1992). Prosthetic patch durability in congenital diaphragmatic hernia: a long-term follow-up study. *J. Pediatr. Surg.,* ***27,*** 754-756.

[29] Urita, Y., Komuro, H., Chen, G., Shinya, M., Saihara, R. and Kaneko, M. (2008). Evaluation of diaphragmatic hernia repair using PLGA mesh-collagen sponge hybrid scaffold: an experimental study in a rat model. *Pediatr. Surg. Int., 24,* 1041-1045.

[30] Peña, A. (1995). Anorectal malformations. *Semin. Pediatr. Surg., 4,*35-47.

[31] Saihara, R., Komuro, H., Urita, Y., Hagiwara, K. and Kaneko, M. (2009). Myoblast transplantation *Pediatr. Surg. Int., 25,* 981-986.

[32] Yiou, R., Yoo, J. J. and Atala, A. (2003). Restoration of functional motor units in a rat model of sphincter injury by muscle precursor cell autografts. *Transplantation, 76,* 1053-1060.

[33] Yiou, R., Dreyfus, P., Chopin, D. K., Abbou, C. C. and Lefaucheur, J. P. (2002). Muscle precursor cell autografting in a murine model of urethral sphincter injury. *BJU Int., 89,* 298-302.

[34] Cannon, T. W., Lee, J. Y., Somogyi, G., Pruchnic, R., Smith, C. P., Huard, J. and Chancellor, M. B. (2003). Improved sphincter contractility after allogeneic muscle-derived progenitor cell injection into the denervated rat urethra. *Urology, 62,* 958-963.

[35] Chermansky, C. J., Tarin, T., Kwon, D. D., Jankowski, R. J., Cannon, T. W., de Groat, W. C., Huard, J. and Chancellor, M. B. (2004). Intraurethral muscle-derived cell injections increase leak point pressure in a rat model of intrinsic sphincter deficiency. *Urology, 63,* 780-785.

[36] Praud, C., Sebe, P., Biérinx, A. S. and Sebille, A. (2007). Improvement of urethral sphincter deficiency in female rats following autologous skeletal muscle myoblasts grafting. *Cell Transplant, 16,* 741-749.

[37] Irintchev, A., Rosenblatt, J. D., Cullen, M. J., Zweyer, M. and Wernig, A. (1998). Ectopic skeletal muscles derived from myoblasts implanted under the skin. *J. Cell Sci. 111,* 3287-3297.

[38] Kasemkijwattana, C., Menetrey, J., Somogyl, G., Moreland, M. S., Fu, F. H., Buranapanitkit, B., Watkins, S. C. and Huard, J. (1998). Development of approaches to improve the healing following muscle contusion._Cell *Transplant, 7,* 585-598.

[39] Hori, T., Giuffra, E., Andersson, L. and Ohkawa, H. (2001). Mapping loci causing susceptibility *J. Pediatr. Surg., 36,* 1370-1374.

[40] Takahashi, K. and Yamanaka, S. (2006). Induction of pluripotent stem cells *Cell, 126,* 663-676.

[41] Yamanaka, S. (2009). A fresh look at iPS cells. *Cell 137,* 13-17.

[42] Sakuragawa, N., Thangavel, R., Mizuguchi, M., Hirasawa, M. and Kamo, I. (1996). Expression of markers for both neuronal and glial cells in human amniotic epithelial cells. *Neurosci. Lett., 209,* 9-12.

[43] Sakuragawa, N., Enosawa, S., Ishii, T., Thangavel, R., Tashiro, T., Okuyama, T. and Suzuki, S. (2000). Human amniotic epithelial cells are promising transgene carriers for allogeneic cell transplantation into liver. *J. Hum. Genet., 45,* 171-176.

[44] Nakajima, T., Enosawa, S., Mitani, T., Li, X. K., Suzuki, S., Amemiya, H., Koiwai, O. and Sakuragawa, N. (2001). Cytological examination of rat amniotic epithelial cells and cell transplantation to the liver. *Cell Transplant, 10,* 423-427.

[45] Miki, T., Lehmann, T., Cai, H., Stolz, D. B. and Strom, S. C. (2005). Stem cell characteristics of amniotic epithelial cells. *Stem Cells, 23,* 1549-1559.

[46] Adinolfi, M., Akle, C. A., McColl, I., Fensom, A. H., Tansley, L., Connolly, P., Hsi, B. L., Faulk, W. P., Travers, P. and Bodmer, W. F. (1982). Expression of HLA antigens., beta 2-microglobulin and enzymes by human amniotic epithelial cells. *Nature, 295,* 325-327.

[47] Sakuragawa, N., Tohyama, J. andYamamoto, H. (1995). Immunostaining of human amniotic epithelial cells; possible use as a transgene carrier in gene therapy for inborn errors of metabolism. *Cell Transplant, 4,* 343-346.

[48] Kubo, M., Sonoda, Y., Muramatsu, R. and Usui, M. (2001). Immunogenicity of human amniotic membrane in experimental xenotransplantation. *Invest. Ophthalmol. Vis. Sci., 42,* 1539-1546.

[49] Sakuragawa, N., Misawa, H., Ohsugi, K., Kakishita, K., Ishii, T., Thangavel, R., Tohyama, J., Elwan, M., Yokoyama, Y., Okuda, O., Arai, H., Ogino, I. and Sato, K. (1997). Evidence for active acetylcholine metabolism in human amniotic epithelial cells, applicable to intracerebral allografting for neurologic disease. *Neurosci. Lett. 232,*53-56.

[50] Elwan, M. A. and Sakuragawa, N. (1997). Evidence for synthesis and release of catecholamines by human amniotic epithelial cells. *Neuroreport, 8,* 3435-3438.

[51] Ishii, T., Ohsugi, K., Nakamura, S., Sato, K., Hashimoto, M., Mikoshiba, K. and Sakuragawa, N. (1999). Gene expression of oligodendrocyte markers in human amniotic epithelial cells using neural cell-type-specific expression system. *Neurosci. Lett., 268,* 131-134.

[52] Kakishita K., Elwan, M. A., Nakao, N., Itakura, T. and Sakuragawa, N. (2000). Human amniotic epithelial cells produce dopamine and survive after implantation into the striatum of a rat model of Parkinson's disease: a potential source of donor for transplantation therapy. *Exp. Neurol., 165,* 27-34.

[53] Uchida, S., Inanaga, Y., Kobayashi, M., Hurukawa, S., Araie, M. and Sakuragawa, N. (2000). Neurotrophic function of conditioned medium from human amniotic epithelial cells. *J. Neurosci. Res., 62,* 585-590.

[54] Sakuragawa, N., Elwan, M. A., Uchida, S., Fujii, T. and Kawashima, K. (2001). Non-neuronal neurotransmitters and neurotrophic factors in amniotic epithelial cells: expression and function in humans and monkey. *Jpn. J. Pharmacol., 85,* 20-23.

[55] Shinya, M., Komuro, H., Saihara, R., Urita, Y., Kaneko, M., Liu, Y. (2010). Neural differentiation potential of rat amniotic epithelial cells. *Fetal Pediatr. Pathol., 29,* 133-143.

[56] Woodbury, D., Reynolds, K. and Black, I. B. (2002). Adult bone marrow stromal stem cells express germlin., ectodermal., endodermal., and mesodermal genes prior to neurogenesis. *J. Neurosci. Res., 69,* 908-917.

[57] Vawda, R., Woodbury, J., Covey, M., Levison, S. W. and Mehmet, H. (2007). Stem cell therapies for perinatal brain injuries. *Semin. Fetal Neonatal. Med., 12,* 259-272.

[58] Sankar, V. and Muthusamy, R. (2003). Role of human amniotic epithelial cell transplantation in spinal cord injury repair research. *Neuroscience, 118,* 11-17.

[59] Okawa, H., Okuda, O., Arai, H., Sakuragawa, N. and Sato, K. (2001). Amniotic epithelial cells transform into neuron-like cells in the ischemic brain. *Neuroreport, 12,* 4003-4007.

[60] Kosuga, M., Sasaki, K., Tanabe, A., Li, X. K., Okawa, H., Ogino, I., Okuda, O., Arai H., Sakuragawa, N., Kamata, Y., Azuma, N., Suzuki, S., Yamada, M. and Okuyama, T. (2001). Engraftment of genetically engineered amniotic epithelial cells corrects lysosomal storage in multiple areas of the brain in mucopolysaccharidosis type VII mice. *Mol. Ther., 3,* 139-148.

[61] Park, W. S., Sung, D. K., Kang, S., Koo, S. H., Kim, Y. J., Lee, J. H., Chang, Y. S. and Lee, M. (2006). Neuroprotective effect of cycloheximide on hypoxic-ischemic brain injury in neonatal rats. *J. Korean Med. Sci., 21,* 337-241.

[62] Erices, A., Conget, P. and Minguell, J. J. (2000). Mesenchymal progenitor cells in human umbilical cord blood. *Br. J. Haematol., 109,* 235-242.

[63] Wang, J. F., Wang, L. J., Wu, Y. F., Xiang, Y., Xie, C. G., Jia, B. B., Harrington, J. and McNiece, I. K. (2004). Mesenchymal stem/progenitor cells in human umbilical cord blood as support for ex vivo expansion of CD34(+) hematopoietic stem cells and for chondrogenic differentiation. *Haematologica. 89,* 837-844.

[64] Mitchell, K. E., Weiss, M. L., Mitchell, B. M., Martin, P., Davis, D., Morales, L., Helwig, B., Beerenstrauch, M., Abou-Easa, K., Hildreth, T., Troyer, D. and Medicetty, S. (2003). Matrix cells from Wharton's jelly form neurons *Stem Cells., 21,* 50-60.

[65] Wang, H. S., Hung, S. C., Peng, S. T., Huang, C. C., Wei, H. M., Guo, Y. J., Fu, Y. S., Lai, M. C. and Chen, C. C. (2004). Mesenchymal stem cells *Stem Cells., 22,* 1330-1337.

[66] Zhang, Y., Li, C., Jiang, X., Zhang, S., Wu, Y., Liu, B., Tang, P. and Mao, N. (2004). Human placenta *Exp. Hematol., 32,* 657-664.

[67] Wulf, G. G., Viereck, V., Hemmerlein, B., Haase, D., Vehmeyer, K., Pukrop, T., Glass, B., Emons, G. and Trümper, L. (2004). Mesengenic progenitor cells derived from human placenta. *Tissue Eng., 10,*1136-1147.

[68] De Coppi, P., Bartsch, G., Jr., Siddiqui, M. M., Xu, T., Santos, C. C., Perin, L., Mostoslavsky, G., Serre, A. C., Snyder, E. Y., Yoo, J. J., Furth, M. E., Soker, S. and Atala, A. (2007). Isolation of amniotic stem cell lines *Nat. Biotechnol., 25,* 100-106.

[69] Sakuragawa, N., Kakinuma, K., Kikuchi, A., Okano, H., Uchida, S., Kamo, I., Kobayashi, M. and Yokoyama, Y. (2004). Human amnion mesenchyme cells express phenotypes of neuroglial progenitor cells. *J. Neurosci. Res., 78,* 208-214.

[70] Tamagawa, T., Oi, S., Ishiwata, I., Ishikawa, H. and Nakamura, Y. (2007). Differentiation of mesenchymal cells derived from human amniotic membranes into hepatocyte-like cells *in vitro*. *Hum. Cell, 20,* 77-84.

[71] Kobayashi, M., Yakuwa, T., Sasaki, K., Sato, K., Kikuchi, A., Kamo, I., Yokoyama, Y. and Sakuragawa, N. (2008). Multilineage potential of side population cells from human amnion mesenchymal layer. *Cell Transplant, 17,* 291-301.

In: Tissue Engineering
Editors: T. Ishikawa and M. Yamamoto
ISBN 978-1-62257-062-1

Chapter 7

TISSUE ENGINEERING STRATEGIES AND STEM CELLS

Markus Tammia, Jonathan Rakar and Joanne M. Hackett*
Department of Clinical and Experimental Medicine,
Linköping University, Linköping, Sweden

1. ABSTRACT

The reoccurring problems faced by the healthcare system today are the loss or failure of tissues or organs. Repairing damaged or diseased tissues or organs, in order to prevent failure using natural or tissue engineered materials, is the essence of regenerative medicine. Technologies can involve the use of stem cells, gene therapy, tissue engineering, and the use of artificial organs. To date, researchers have been recreating a range of tissues and organs in vitro, with varying degrees of success. The tools to create scaffolds and structures are limited, and fine control over mechanical and environmental variables is complex. Also, there is a limited means of diagnosing at the molecular, cellular, and tissue levels what is actually happening within the constructs. In this chapter, we focus on advancements in different areas of tissue engineering and regenerative medicine starting with the importance of scaffolds for tissue engineering. Further on we describe strategies for nerve regeneration both in the PNS and CNS and end in a brief review of sources of multipotent cells for tissue engineering, including remarks on the use of genetic techniques for in vitro differentiation.

2. INTRODUCTION

"We are all sculptors and painters, and our material is our own flesh and blood and bones" - Walden is still mostly interpreted in a metaphorical rather than in a literal (clinical)

* Corresponding author: P: +46-(0)76-394 6905; F: +46-(0)10-103 4149. E-mail: Joanne.Hackett@liu.se.

way. The optimism surrounding the advent of tissue engineering, itself an ancient concept, is motivated by the excitement carried by a more literal understanding of those words of Thoreau [1]. With the development of human cell cultures comes the ability to give back more cells than were taken from a biopsy, effectively amplifying the cell numbers available for autografts. With the realization that cells could not only be multiplied but also controlled by their culture environment spawned hope of using cellular engineering approaches to recreate damaged tissue or entire organs - to build perfect biological substitutes. The junction between molecular biology, cellular biology, and materials science is where the study of cells, biochemical factors and materials together set the stage for tissue engineering, leading us ever closer to using cells as building blocks for human replacement parts. The term "tissue engineering" was coined at a National Science Foundation committee meeting in 1987 [2] but the promise of a fully functional engineered tissue is still unkept today.

The lack of major clinical breakthroughs should not be interpreted as a lack of progress within tissue engineering. Increased understanding of the complexity of the field has deepened the focus of research, and perhaps this has humbled a generation of prospective tissue engineering pioneers. It was only in the 1960's that knowledge of a finite cellular life-span was obtained through the research of Hayflick [3] and still today, building on work from 1957 by Waddington, the concepts of epigenetics are not completely understood [4]. Considering the progress still to be made in basic cell biology, it is not hard to fathom that full control of cell cultures for tissue engineering needs a few years yet to be successful. Much effort is exerted toward that goal; not only is it pertinent to maintain healthy and stable cultures even under extensive expansion phases, but it is also desirable to control the phenotypic identity of the cells in culture - either to maintain them or to allow them to differentiate.

3. Bioengineering Strategies Used in Regenerative Medicine

Within regenerative medicine, bioengineering aims to create materials that support and assist the human body's own regenerative potential. This can be achieved using materials that provide the healing tissue with proper guidance and a beneficial environment, using mechanically and environmentally supportive constructs and targeted drug, cell or gene delivery.

3.1. The Extracellular Matrix as a Gold Standard

The extracellular matrix (ECM) is the framework in which cells live and proliferate in native tissue. It provides a dynamic, highly porous network of nonwoven fibers lined with ligands that convey signals determining cell fate and attachment. Through cell-surface receptors, such as integrins, the cells bind to the ligands that are presented by the ECM. This promotes a signal that is further transduced into chemical signaling cascades within the cells. In this way, cells can sense biophysical queues from the ECM that determine cell fate, *e.g.* migration, proliferation or differentiation [5]. The ligands of the ECM work in synergy and the spatial arrangements of them influence cell behavior. For instance, a 4 nm colocalization between the synergic site Pro-His-Ser-Arg-Asn (PHSRN) and the integrin binding sequence Arg-Gly-Asp (RGD) induces increased osteoblast metabolic activity and alkaline phosphatase production while reducing the ECM production. Through differences in the topography and ECM protein presentation, each type of tissue tailors the response of the cells within it [6, 7].

3.2. Cues that Influence Cell Response

In addition to matrix or scaffold topography and ligand presentation, the mechanical properties are important to consider in the design of scaffolds. In addition to providing sufficient mechanical support for the healing tissue, the stiffness of the substrate needs to promote cell growth and appropriate phenotype expression. Materials with elastic moduli comparable to the native tissue have been shown to promote cell differentiation and proliferation for a number of tissues [8-10], and can be crucial to maintain stem cell regenerative potential when cultured *in vitro* [11]. Furthermore, it has been indicated that matrix compliance modulates the spatial presentation of ECM ligands to the cells [9].

To mimic the qualities of the ECM, bioengineering currently aims to design bioactive three dimensional (3D) scaffolds. Compared to planar surfaces, or 3D networks of large dimensions, 3D scaffolds in a size range comparable to that of the ECM will display a more natural cell response [6]. The dimensions effect on cell differentiation was demonstrated by Christopherson *et al.* in a study where they investigated rat neural stem cell (NSC) response under different culturing conditions to nanofibrous scaffolds with varying fiber diameters. They were able to show that different fiber diameters favored the differentiation into different lineages [12].

3.3. Material Construction and Design

Implants can be constructed to be either biodegradable or not, depending on the desired mode of action. Nondegradable implants have the benefit of not having to account for issues related to degradation products to the same extent as degradable materials. They are also most often made out of synthetic materials. These characteristics together provide for a more simple design and more precise control over the material properties. Since these materials do not necessarily accommodate healing tissue, this could imply the need of a further intervention to remove the implant when the need of its support has expired. This can be avoided through the use of degradable materials. The concept is that the scaffold should provide a temporary support for the healing tissue, and make way for regenerating natural tissue through degradation. In order for this to work, the rate of degradation must be tuned so that it provides sufficient support over the regeneration process, while not hindering tissue repair. The biomaterial must also degrade in a way that does not leave byproducts detrimental to the healing process, as in the case for poly(α-hydroxy acids) where the degradation leads to a decline in pH, which can inflict cell death if the degradation is too rapid. Biodegradable scaffolds are comprised of either synthetic polymers, *e.g.* poly(glycolic acid) (PGA), or natural polymers, such as collagen or chitosan. When materials harvested from a natural source are used, there is always a certain batch-to-batch variability that can restrict the control over the material properties. Furthermore, insufficient purification can leave residues that evoke an immune response to the implant. The benefit of using natural polymers is that they can present easily degradable substrates with degradation products common to the body. They also benefit from their inherent ligands to which cells can attach [13].

Since many synthetic materials lack the ligands for cell attachment naturally found in materials such as fibronectin, collagen, laminin or chitosan, their surfaces can be treated to assist cell attachment. This can be achieved with a surface coating by a protein, *e.g.* laminin, or by merely incorporating attachment motifs or molecules, such as the common RGD motif or terminally aminated branches, respectively. By variations in the ligands used, cell differentiation on the material can be altered [13, 14]. The presentation of the ligands to the cells will alter their influence on cell attachment and fate, as was shown in the studies by Chua *et al.* [15, 16]. They showed that aminated nanofiber scaffolds had better expansion potential for human umbilical cord derived hematopoetic stem/progenitor cells (HSPCs) over carboxylated or hydroxylated scaffolds. Further studies showed that the use of spacer molecules enhanced the cell adhesion and that the length of the spacer molecule had an impact on cell gene expression. The use of a spacer is thought to make the ligand more accessible to cells by reducing sterical hindrance of ligand-receptor interactions.

When in an *in vivo* application or in a serum based medium, one has to consider protein attachment to the scaffolds. When approaching the scaffold, the cells will not directly sense the actual scaffold surface, but rather the adsorbed protein layer present [17]. For protein adsorption, the hydrophilicity of the material is important. This was demonstrated in a study by Lee *et al.* who found that PC-12 cell adhesion coincided with improved nerve growth factor and serum protein adsorption to poly(ethylene) substrates of intermediate wettability. Although the cell adhesion was best to the surfaces of intermediate wettability, a more hydrophilic nature gave enhanced differentiation into neuron like cells [18]. Li *et al.* further highlighted the impact of surface chemistry and wettability on cell attachment. They prepared polymeric substrates with grafted poly(acrylic acid) to form a gradient of carboxyl groups. Some of the substrates were then treated by argon plasma, which introduces oxygen containing groups to the surfaces, thus increasing the wettability. The C17.2 cell attachment and differentiation was dependent on carboxyl group density. The difference in cell attachment characteristics was overcome by the argon plasma treatment, resulting in uniform cell distribution over the gradient [19].

3.4. Biomaterials and Stem Cell Based Therapies

Bioengineered materials and stem cell based therapies go hand in hand within regenerative medicine. Stem cells or progenitor cells have the ability to differentiate into several different cell types. Therefore, they serve as a base for reestablishing the damaged tissue. Biomaterial approaches can be utilized to either recruit host cells through chemical cues or as a vehicle for cell delivery [20, 21].

As for cell delivery, injectable hydrogels are often employed vehicles. Hydrogels are networks of highly hydrated polymer chains that are held together as a gel phase through either chemical crosslinking or by weaker interactions such as physical entanglement, electrostatic interactions, hydrogen bonds or crystalline segments. A so called sol-gel transition occurs as the polymers in the suspension interact and get entangled to a degree that the solution goes from a liquid to a gel state [22]. Through this, *in situ* gelation can be triggered by changes in temperature, pH, salinity, *etc.* [21]. Injectable hydrogels bring along several interesting features as materials for tissue regeneration. They can be used as vehicles for cell and drug delivery and can be injected as viscous solutions to precisely fit the site of the injury after gelation [21, 22]. Several of the conditions under which the sol-gel transitions take place are too harsh for delivering certain drugs or living cells. This can be circumvented by appropriate design strategies of physical hydrogels, as recently demonstrated by Foo *et al.* They utilized a protein based hydrogel that gelated as two components were mutually injected. These mild gelation conditions were proven feasible for *in situ* cell delivery [21].

Nanofibrous scaffolds to mimic ECM architecture can be made through hydrogel self assembly, in which weak, non-covalent interactions work in concert to build up organized structures from the bottom up. The methods make use of dispersion forces, such as induced dipoles and Van der Waal interactions, hydrogen bonds and hydrophobic/hydrophilic interactions to build up supramolecular structures. The interactions employed have a more or less determined spatial restriction, which also is displayed in the body's own ligand-receptor complexes, and is used as spatial cues for the assembly process. To govern the formation of these hierarchical structures, conditions such as pH and ionic strength are tuned to favor the desired arrangements. Some of the problems with self assembled nanofibrous scaffolds are difficulties scaling up the solid-phase peptide synthesis, control over degradation rate and mechanical weakness of the hydrogels, limiting them from transferring mechanical loading to cells [5, 23].

3.5. Fabrication Techniques

In addition to hydrogels, there are a number of other techniques that try to mimic the 3D structure of the ECM. Lately, the technique of electrospinning has become increasingly popular. In electrospinning, polymer fibers are formed from polymer solutions or melts using a high voltage. Through this, fibers with diameters in the micro- to nanometer range can be collected either in oriented, aligned or random meshes with highly interconnected pores. The porous structure in this size range provides cells with the topological cues discussed above, and allows for effective flow and diffusion of nutrients and waste products throughout the scaffold. There is a vast range of polymers suitable for electrospinning, with a continuous development adding new materials to the list. The process also benefits from the fact that blends of polymers and other substances can be spun into the fibers, although one has to account for the impact that the spinning conditions might have on the activity of the substances [24-27].

Similar fibers to those produced by electrospinning can be produced by a technique called rotary jet spinning, as demonstrated by Badrossamay *et al.* [28]. Here, aligned fibers are extruded by using a high speed rotating nozzle fed with a polymer solution. This technique does not employ high voltage, adding to the types of fibers that can be produced in this size range, some of which might not be attainable by electrospinning.

Within hard-tissue engineering, the challenge is to find materials that provide for a better healing response than seen with traditional materials. With the use of traditional metals and ceramics, a hard capsule is formed around the implant, preventing regeneration. To stimulate bone regeneration, mesoporous nanocomposites and bioceramics have attracted the interest of researchers, where polymers are blended with biological components [5, 29].

By their ordered porous structure, they can provide a large surface area on which they present nucleation sites for bone mineral deposition, and in this way enhance the bioactivity rate [30].

3.6. Conclusion

In conclusion, the focus in bioengineered materials is to provide a support in the form of an artificial ECM or cell niche. The design strategies employed and investigated draw from the increased understanding of the interactions cells have with the materials, regarding chemistry, mechanical properties, spatial dimensions and architecture.

4. Nerve Regeneration

4.1. The Peripheral Nervous System (PNS)

4.1.1. Introduction to the PNS

The portion of the nervous system that is outside the brain and spinal cord is referred to as the peripheral nervous system (PNS). The nerves in the PNS connect the central nervous system (CNS) to sensory organs (such as the eye and ear), other organs of the body, muscles, blood vessels and glands. The peripheral nerves include the 12 cranial nerves, the spinal nerves and roots, and what are called the autonomic nerves. The autonomic nerves are concerned with automatic functions of the body. Specifically, autonomic nerves are involved with the regulation of the heart muscle, the tiny muscles lining the walls of blood vessels, and glands.

Peripheral nerve damage can lead to pain, tingling, numbness or difficulty coordinating hands, feet, arms or legs. This can happen with diseases like diabetes, an injury due to a crushed or cut nerve, or other conditions known as neuropathy. Surgical restoration of function following peripheral nerve injury has made significant progress since 1879 [31]. Advances in microscopy and microsurgical techniques, and an enhanced comprehension of the internal topography of peripheral nerves, has greatly improved functional outcomes.

The ultimate goal of any peripheral nerve reconstruction is the restoration of function as fully and quickly as possible, while minimizing donor site and systemic morbidity.

Table 1. Summary of surgical restoration options following peripheral nerve injury

Type of repair	Example
Tension-free	Primary end-to-end neurorrhapy
Nerve graft	Autogenous sural nerve graft
Nerve transfer	Distal spinal accessory to the suprascapular nerve
End-to-side neurorrhapy	Brachial plexus
Allografts	ABO matched donors for small diameter donor nerves
Conduits	Polyglycolide as a nerve repair wrap for noncritical small diameter sensory nerves

In cases where a tension-free primary end-to-end neurorrhaphy is not possible, several alternatives exist. This section summarizes these options for repair including interpositional nerve grafting, transfers and end-to-side neurorrhaphy (Table 1).

4.1.2. Tension-Free Primary Repair Alternatives

When tension-free primary repair is not possible, a suitable alternative must be pursued. The surgical technique employed in these alternatives is similar, whether it be a nerve graft, conduit, or nerve transfer. The proximal and distal nerve ends are trimmed until normal fascicular structure is revealed, allowing the surgeon to achieve appropriate realignment.

Nerve Grafting

Autogenous nerve grafting has long been considered the "gold standard" for repair of irreducible nerve gaps. Autogenous grafts act as immunogenically inert scaffolds, providing appropriate neurotrophic factors and viable Schwann cells (SCs) for axonal regeneration. The choice of autogenous graft is dependent on several factors: the size of the nerve gap, location of proposed nerve repair, and associated donor-site morbidity. Although the sural nerve is the most commonly used autograft, there are many other suitable nerves that can be used as interposition grafts including: the medial and lateral cutaneous nerves of the forearm, dorsal cutaneous branch of the ulnar nerve, superficial and deep peroneal nerves, intercostal nerves, and the posterior and lateral cutaneous nerves of the thigh [32, 33]. As with all repair techniques, when utilizing an interposition graft it is critical to avoid any tension at the repair site, for even minimal tension can negatively impact the functional outcome [34]. The advantages of this technique include the fact that the nerve gets bridged using a nonimmunogenic medical device. The disadvantages, however, include sensory loss, scarring, neuroma formation, second incision, limited supply, and the fact that it is inferior to tension-free primary repair.

Nerve Transfers

Nerve transfers convert a proximal injury into a distal one by transferring "near by" redundant nerve function to a distal denervated nerve close to the target. The concept of nerve transfer itself is not new [35, 36], but has gained significant momentum over the past decade with many authors reporting good outcomes [37-40]. The use of nerve transfers for proximal upper extremity and brachial plexus reconstruction is now well established. With all transfers, when selecting the appropriate donor nerves several criteria must be considered: proximity of donor nerve to the recipient nerve motor end plates, availability of redundant or expendable donor nerve, synergism of donor muscles to target muscles, similarity in number of motor or sensory axons between donor and recipient nerves, and size matching. One advantage of this technique is one typically avoids donor autografts and associated morbidity, as well, proximity of donor nerves to target motor endplates provides earlier reinnervation. The disadvantages include possible loss of function from the donor nerve site, and that the donor muscle is no longer suitable for muscle transfer.

End-to-Side Neurorrhaphy

Described over a century ago [41, 42] end-to-side coaptation is an alternative in cases where the proximal nerve stump is not available or inaccessible; instead, the injured distal stump is coapted to the side of an uninjured donor nerve. In 1994, Viterbo *et al.* sparked new interest in this technique by reporting axonal regeneration with the use of end-to-side neurorrhaphy in a rat model [43]. This revitalized interest has given rise to a substantial amount laboratory and clinical work investigating its application [44-49]. Although there are mixed reviews [50, 51], the controversy surrounding end-to-side has primarily focused on the requirement for donor nerve axotomy. In an end-to-side repair, the term "collateral sprouting" refers to novel axonal regeneration into the recipient stump originating from the donor nerve; the exact origin of these new axons is disputed. Hayashi *et al.* designed an end-to-side coaptation rat model that left the donor nerve intact. Evidence of collateral sprouting was initially reported even in the absence of donor nerve injury [52]. However, when repeated in a more sophisticated model, it was shown that only sensory axons sprouted *de novo* without injury [53]. Matsumoto *et al.* also demonstrated that although collateral sprouting occurs in the absence of donor nerve axotomy, only sensory reinnervation occurs [54]. Therefore, it is believed that in order for motor fibers to regenerate, a donor nerve axotomy must be performed. This technique is useful when minimal sensory recovery is needed, however there will be limited to no motor recovery without a donor axotomy.

Nerve Allografts and Conduits

The use of donor related or cadaveric nerve allografts is reserved for devastating or segmental nerve injuries. Like all tissue allotransplantation, nerve allografts require systemic immunosuppression; the associated morbidity of immunomodulatory therapy limits the widespread application of nerve allografting. Several techniques (*e.g.* cold preservation, irradiation, lyophilization) to reduce nerve allograft antigenicity have been described [55, 56]. In contrast to autografts, the use of nerve allografts relies upon the viability of both host and donor SCs. The donor SCs play a dual role following nerve transplantation, acting as support cells for remyelination and as facultative antigen presenting cells. This dual role prohibits a complete removal of these presumed primary sources of major histocompatibility complex (MHC) II molecules and simultaneously maintains the requirement for systemic immunosuppression [57, 58]. Unlike solid organ transplantation, nerve allotransplantation requires only temporary immunosuppression. Once adequate host SC migration has occurred into the nerve allograft at approximately 24 months, systemic immunosuppression can be withdrawn. Current research efforts are aimed at understanding the host response to nerve allotransplantation and mechanisms to modify host and/or donor alloantigen recognition and presentation. This understanding may lead to further refinement of the current immunomodulation strategies and may eventually provide an unlimited supply of nerve graft material. Allografting is readily accessible, in an unlimited supply, it bridges the nerve gap, and avoids donor site morbidity. The disadvantage is the potential side effects of host immunosuppression.

To avoid donor site morbidity, a viable synthetic or biologic nerve conduit is an option. The use of nerve conduits is often restricted to the repair of noncritical small diameter sensory nerves, gaps less than 3 cm, and as a nerve repair wrap. The concentration of neurotrophic factors appears to be critical to advancing nerve regeneration, and if the volume of a conduit increases beyond a critical size, regeneration is inhibited unless the length of the conduit is dramatically shortened. As surgeons achieve clinical success with short-gap, small diameter nerve injuries, the clinical front is being extended to include larger diameter nerves at longer gaps [59]. Conduits are readily available, avoid donor site morbidity, bridge a nerve gap, acts as a barrier to scar tissue infiltration, and may allow for accumulation of local neurotrophic factors. These conduits, however, seem to be limited to short nerve gaps, and achieve variable outcomes.

4.1.3. Conclusion

Regeneration in the PNS offers unique challenges and opportunities to clinical and translational neuroscience. While each repair strategy discussed above has its own advantages and disadvantages, careful consideration of outcome goals and individual patient factors usually affords a directed surgical approach. The appropriate intervention is ultimately determined by injury type, patient characteristics, and surgeon preferences.

4.2. The Central Nervous System (CNS)

4.2.1. Introduction to the CNS

While the peripheral nervous system possesses innate regenerative capacity, partly attributed to the presence of Schwann cells, there is no sufficient treatment for injuries to the central nervous system in general, and to the spinal cord in particular [13]. In this section, the pathophysiology of spinal cord injuries will be reviewed, from the initial trauma that then extends into a secondary injury and the formation of a glial scar that has been found to inhibit axonal regrowth. The functions of the constituents that orchestrate the injury development are not fully understood, hence there is no clear consensus as how a successful regeneration would be achieved.

The initial mechanical injury can be a fracture or a dislocation of the spinal column, causing damage to axons, blood vessels and cell membranes. The blood brain barrier (BBB) is disrupted and its integrity lost. The degree to which the spinal cord is damaged has a great impact on the outcome. If a full injury is inflicted, function is lost from below the point of injury. In contrast, a preservation of as little as 5% of the axons has been found capable of sustaining substantial neurological functions [60].

After the primary injury, a prolonged period of continued tissue damage called the secondary injury occurs. The spinal cord begins to swell immediately after the injury, with hemorrhaging in both the white and the central grey matter, leading to ischemia and necrotic cell death that extends both rostrally and caudally from the site of the injury [60].

To help clear the debris at the site of the injury, immune cells are recruited. While doing so, they inflict collateral damage to the nearby cells, thus leading the injury to extend. Immediately after the injury, microglial cells are activated. To make the BBB more permeable for recruited inflammatory cells, tumor necrosis factor-α (TNF-α) and interleukin 1-β (IL-β) are upregulated and matrix metalloproteinases (MMP) are released [60, 61]. The extent to which IL-β is expressed has been found to correlate with the severity of the injury, while observations have been made of this correlating with protective characteristics through increased expression of trophic factors [62]. MMP-9 which is expressed by neutrophils increases BBB permeability by degrading some of its components. In doing so, it also demyelinates uninjured axons, leaving them exposed and more vulnerable. However, MMP-9 could serve a beneficial role by breaking down the glial scar forming at later stages of the injury [61].

The processes of the secondary injury become dominant as the acute stage sets in at two hours to two days post injury. Immune cells invade the site of the injury, bringing with them a vast amount of non-cellular mediators. The axons take further damage from neurotoxicity mediated by the immune response and the presence of free radicals, as well as glutamate excitotoxicity and ionic dysregulation [60]. Dying cells can no longer maintain the extracellular environment due to failure of their energy dependent transporters [63]. Oligodendrocytes tend to undergo apoptosis, which has been linked to ligands expressed on activated microglial cells, signaling through the p75 neurotrophin receptor. Oligodendrocytes are also susceptible to the glutamate excotoxicity due to their expression of N-methyl-D-aspartic acid (NMDA) receptors. This death of oligodendrocytes further demyelinates the axons [60].

Using enzymes, neutrophils try to clear out the debris. They appear around four to six hours after the trauma, with a peaking response after one to two days. Clearing the area they also inflict damage to adjacent cells. Limiting the neutrophil influx has been shown to reduce the secondary injury to neurons in animal models [61]. Another way of restricting the extent of the secondary injury could be to use antioxidants to counter the oxidation of cell membrane lipids by free radicals [60].

When these events subside, an elevated phagocytic activity clearing debris marks the onset of the subacute stage between two days and two weeks after the injury. It is during the subacute stage when the immune mediated neurotoxicity has been dampened that cell based therapies show the most promise [60]. The cells recruited for the phagocytosis are macrophages derived both from blood borne monocytes and resident microglial cells. Even though the macrophages clear debris that contains moieties inhibiting axonal regeneration, they are considered to inflict severe damage to the nerve cells through the pro-inflammatory cytokines, reactive oxygen species, nitric oxide and proteases that they release [61]. Studies have also indicated that macrophages cause astrocyte migration away from them, making neurons lose their support and take damage from stretching of the axons as the astrocytes migrate. Due to this, it is believed that the severity of the secondary injury can be reduced by restricting early immune cell infiltration [64].

During the subacute stage, the so called glial scar starts to form. It consists of astrocytes, oligodendrocytes, microglial and precursor cells and their secreted proteins [60, 63]. The scar is formed by astrocytes at the periphery of the injury that become hypertrophic and proliferative, creating a physical and chemical barrier for axonal growth with an enrichment of inhibitory myelin proteins [60, 65]. The astrocytes enrich at the site through both migration and proliferation [66]. Meningeal cells also invade the lesion, further enhancing the scarring by interacting with the astrocytes [67]. The reactive astrocytes do however serve beneficial functions to the healing as well; they promote the reestablishment of ionic homeostasis and reconstruct the BBB which in turn limits the edema and invasion of activated immune cells [60]. A removal of reactive astrocytes could promote axonal regeneration, but it would impair the restoration of the BBB. If the BBB fails to reestablish, it would turn the secondary injury into a long lasting inflammation [62, 66].

Angiogenesis is an important part of the healing process, active until at least 28 days after the injury. This is thanks to upregulated growth factors and chemokines. In rat models of spinal cord injuries, angiogenesis has been found to occur in two bursts, the first one correlated with regenerative and behavioral improvement [65].

The subacute stage is followed by the intermediate stage with a continued maturation of the glial scar between two weeks and six months after the injury. There is now some regenerative axonal sprouting [60]. A mature scar is achieved at the start of the chronic stage and Wallarian degradation breaks down distal parts of damaged axons [60, 68]. Cysts and syrinxes are formed and the scar is fully matured after one-two years with cystic cavities and myomalacia, the softening of the spinal cord. There might still be some delayed neurological dysfunctions and neuropathic pain due to syrinx formations [60].

4.2.2. Repair and Regeneration Strategies

Since peripheral nerves are more permissive to regeneration, methods to translate these attributes to the CNS have been sought. Attempts in using peripheral nerve grafts have elicited axonal regeneration, but this has been dependent on the axon type. While the axons have been able to enter the graft, success in reentering the CNS has still been limited [69, 70]. Another promising approach in translating regenerative attributes to the CNS is the use of transplanted Schwann cells. These cells have been thought to account for the regenerative capabilities of the PNS through clearing out debris and creating a beneficial, protective environment for regenerative axons. The Schwann cells do this through secretion of neurotrophins and presenting cell adhesion molecules, laminin and fibronectin, thus guiding and promoting axonal regeneration. Observations that Schwann cells migrate into injuries in the CNS have been made, but their role is still unclear. Transplanted Schwann cells have been successful in transferring regenerative properties to the graft and remyelinating axons. Still, these grafts have been unable to permit regenerating axons to exit the graft [13, 70].

A variety of cell sources have been investigated for cell based therapies. These therapies work through a variety of ways; they serve a base for cellular replacement for both remyelination and neuronal replacement, provide an environment permissive to axonal regeneration and serve to protect the neurons. By secreting growth factors, transplanted cells alter the milieu at the injury site, potentially promoting regeneration [69-72].

Olfactory ensheathing glia (OEG) is a cell type present in the olfactory bulbs, the only part of the CNS known to retain embryonic properties throughout life. In the olfactory bulbs, OEGs form channels that enable severed axons to reestablish the lost connections. Due to these properties, OEGs have been transplanted to the injured spinal cord. Here, OEGs have been shown to guide regenerating axons past the glial scar. A possible explanation could be that the cells give a reduced expression of inhibitory molecules on the host astrocytes. The cells also promote vascularization and remyelination, even though they themselves cannot provide myelination of the axons. The OEGs can be transplanted as purified cells, or co-transplanted with olfactory nerve fibroblasts, with which they have an intimate interplay in the olfactory. These cells together build up a matrix that promotes axonal regeneration [69-71].

Stem cell therapies have been sought to replace the injured cells as they can differentiate into all lineages required in the injured spinal cord. Results may however vary depending on cell source and how they have been treated before engraftment. Embryonic stem cells (ESC) have been shown to remyelinate axons in myelin deficient mice, so called *shiverer* mice. In order for this to work, the cells need to be differentiated into glial precursor cells before seeding [71]. This can be done with high efficacy and in great numbers, but ESC derived cells face both ethical issues regarding the use of embryonic material, as well as the need for immunosupression since non-autologous cells are used [72]. The recently developed technique of inducing embryonic characteristics to adult cells, induced pluripotent stem cells (iPSCs), could provide patient specific stem cell therapies [73, 74]. There is however always a concern for tumor formations when cells with embryonic stem cell properties are utilized, although this has yet not been reported for ESC derived oligodendrocyte precursors [72]. Neuronal stem cells (NSCs) are an alternative that also have remyelinating potential and can replace damaged neurons. However, neuronal and oligodendrocyte differentiation is restricted in SCI, with a bias towards a glial phenotype. This stem cell behavior differs from *in vitro* studies, indicating that there are factors present at the site of the injury restricting neuronal differentiation. To overcome this hurdle, methods that pre differentiate the cells into neuronal lineages, or create an environment promoting neuronal differentiation are sought [71]. Since the propagation of adult NSCs is restricted, the need for fetal NSCs immediately poses ethical restrictions to their use [72].

In the adult spinal cord, oligodendrocyte precursors, ependymal cells and astrocytes are the cell types that have displayed stem cell properties. A recently published study compared the reproduction rate and progeny of these cell types in both healthy and injured mice. In the uninjured spinal cord, ependymal cells and astrocytes maintained their population and did not give rise to progeny of other lineages, while oligodendrocyte precursors both self renewed and produced new oligodendrocytes. When SCI was inflicted, astrocytes and ependymal cells had a high increase in production of progeny. Cells derived from all three cell types were found at the lesion, but in complementary domains with astrocytes progeny at the periphery. Amongst the three cell types investigated, only astrocytes and ependymal cells produced progeny positive for nestin, marking scar forming astrocytes. The absolute majority of remyelinating oligodendrocytes were originating from oligodendrocyte precursor cells. While ependymal cells can produce oligodendrocytes, most of their progeny was made up of astrocytes. The authors proposed that a shift of ependymal cell differentiation towards the oligodendrocyte lineage could lessen the extent of glial scarring while promoting remyelination [75]. To use recruitment of an endogenous stem cell population for cell replacement could enable minimally invasive therapies. In order for this to prove beneficial and not to increase the scar formation the cell fate must be controlled. Through studies of transplanted cells, it has been shown that the cell niche at the site of transplantation has a great impact on stem cell fate. Methods to create an environment that favors oligodendrocytic and neuronal differentiation for axon regeneration and remyelination with a lessened glial scar must be investigated. It is however unclear if a weakened astrocyte differentiation would be able to reestablish adequate tissue integrity [72].

4.2.3. Biomimetic Alternatives

In spinal cord regeneration, materials have been employed to bridge the gap of the injury, creating a support for the healing cord. The materials need to meet a variety of demands for successful healing. One of the most basic requirements is to have mechanical properties that can cope with the movement of the spinal cord. The implant must also have mechanical and chemical properties that allow for proper cell response, *i.e.* to promote the healing and favor the attachment, survival and proliferation of CNS cells [6, 13, 76]. The porosity of the material must support effective diffusion of nutrients and waste products, while not restricting cell migration [76]. To meet these complex demands, a variety of the techniques described earlier in this chapter have been employed. A combination of complementary techniques is likely required to simultaneously meet the requirements.

There have been two main design strategies for spinal cord regeneration scaffolds; tubular scaffolds and injectable hydrogels. A tubular scaffold can be employed either as unfilled scaffolds [77] or with a filling compound [78]. This can be constituted of smaller guiding tubes or internal matrices made from fibrous networks or hydrogels. By this approach, the different compounds of the scaffold serve to meet different parts of the design requirements. An external tube can provide a greater mechanical and structural support, while the internal matrix accounts for cell fate and gap filling properties, and can be designed to deliver cells or active compounds such as growth factors. The guidance channel approach has the drawback that its implantation is very invasive. This can be avoided through the use of an injectable hydrogel. An injectable hydrogel is also very well apt to cope with variations in the shape of the lesion [13]. However, the design becomes increasingly complex as the material alone needs to meet all the requirements, and compromises are seemingly inevitable. Mechanical properties and surface topography has been given an increasing attention in tissue engineering in general. For CNS applications in particular, these properties have been shown to have an important impact in determining cell fate. While neurons prefer softer substrates, a slightly more rigid material promotes astrocytes [79-82]. Christopherson *et al.* [12] also demonstrated that adult rat neural stem/progenitor cells are sensitive to substrate dimensions in fibrous scaffolds. For their materials, an average fiber diameter of 283 nm favored oligodendrocytes differentiation, while an average fiber diameter of 749 nm promoted neural differentiation. The importance of successful cell attachment was also demonstrated as an average fiber diameter of 1452 nm led to decreased cell attachment coupled with a reduced viability. Lim *et al.* [83] later demonstrated that fiber alignment also serves a selective purpose for these cells. Using neural differentiation inducing medium, aligned fibers gave a more efficient differentiation compared to randomly oriented meshes and planar surfaces, while reducing oligodendrocytes attachment and survival. As discussed previously in this chapter, hydrogels can serve as vehicles for delivery of both cells and biomolecules. In contrast to direct injection, hydrogels serve to localize the delivery to the site and to protect the molecules from degradation after injection. To gain regeneration, one can either stimulate cells through neurotrophic factors to rebuild damaged tissue, or battle the inhibitory effects of the glial scar through antagonists against inhibitory moieties or enzymatic degradation of the inhibitory molecules. The neurotrophic factors can be delivered in a variety of ways; cell induced release of molecules coupled to the scaffold, gene delivery to make endogenous cells produce neurotrophins, controlled release from polymeric microcarriers, or by encapsulating cells secreting neurotrophins, such as those previously discussed. Methods for a controlled temporal and spatial release are being investigated [13].

5. Multipotent Cells in Tissue Engineering

5.1. Introduction

The realization that blood cells were continually replenished throughout life spurred the discovery of the hematopoietic stem cells of the bone marrow. This eventually lead to attempts in the 1950's to treat radiation damage using allogeneic bone marrow transplantation - a strategy that became immensely useful together with the understanding of the major histocompatibility complex within the following two decades. Moreover, bone marrow stromal cells that support development of cell fates outside of the hematopoietic lineage were also discovered in the marrow microenvironment. Stem cells are considered immensely important for tissue engineering owing to their ability to yield a number of different specialized cells upon controlled stimulation *in vitro*. This ability to differentiate has been shown to various degrees for different stem cell populations, and classifications range from the totipotent fertilized egg that can create the entire human, including the extraembryonic tissues, through pluripotent and multipotent stem cells down to unipotent progenitor cells.

The embryonic stem cells, often obtained from the inner cell mass of the blastocyst, show a high degree of stem cell plasticity but at the cost of phenotypic volatility. Embryonic stem cell cultures are difficult to control and are hampered by ethical misgivings, public misconceptions and non-harmonized global legislation. Additionally, these cells are never an autologous source, limiting their use in tissue engineering. In contrast, adult stem cells generally demonstrate lesser degrees of phenotypic plasticity making them versatile yet controllable. They can be obtained from a number of anatomic locations of any adult making them relatively available, potentially autologous and hardly controversial. Such a brief summary as this introduction may spawn the idea that the choice of cells in tissue engineering is rather simple. That is not the case.

5.2. Sources and Functions of Multipotent Cells

In the 1960's, Friedenstein [84] used fibroblast colony forming assays (CFU-F) to investigate the nature of bone marrow fibroblasts and their capacity for microenvironment maintenance and differentiative capacity. Friedenstein recognized the marrow stromal cells as a heterogeneous fibroblast population with an important role in supporting hematopoiesis, which he believed proved the hypothesis of a stromal-hematopoietic relationship dictating hematopoiesis as proposed by Alexander Maximov (1874-1928) [85]. Besides a growing understanding of the surface antigen profiles of multipotent marrow stromal populations Friedenstein's research also showed the importance of growth factors such as TGF, EGF, FGF and PDGF for maintenance and differentiation of multipotent cells in vitro [86]. Some consider the multipotent marrow stromal cells, often referred to simply as mesenchymal stem cells, as the most potent adult stem cells and therefore the most suitable for tissue engineering.

The advent of flow cytometry has enabled the selection of cells in suspension *en masse* according to their surface antigen profiles, which has proven useful for characterizing and isolating cell populations. Pittenger and co-workers [87] used flow assisted cell sorting to isolate human mesenchymal stem cells based on a specific selection of CD markers (CD29, CD44, CD71, CD90, CD106, CD120a and CD124 and negative selection for hematopoietic stem cell markers CD14, CD34, CD45). This population was shown to be multipotent by showing *in vitro* differentiation to adipogenic, osteogenic and chondrogenic lineages upon defined medium supplementation. Clonal colonies used in differentiation experiments proved to be uni-, bi- and tripotent despite being indistinguishable in terms of marker profile and marker stability. This lead to speculation that the cultures may have acquired genomic alterations after 21 population doublings, or that subsets of cells were variably committed progenitor cells [87]. These results exemplify an important problem in stem cell biology, which is the lack of a full biological understanding of what comprises a specific phenotype. This limitation leads to approximations when defining any population, and concomitantly leads to difficulties in determining the degree of success when stem cells are differentiated. Accordingly, since the publication of Pittenger *et al.* (1999) further characterizations of marrow sub-populations have been published, but these will not be described here [briefly reviewed in [88]].

Many other stromal compartments have been found to host populations of multipotent cells, such as the liver, lungs, dental pulp, thymus, and even such anatomical sites that were thought to be relatively lacking in regenerative capacity (heart and brain). In the heart, resident primitive cells (progenitors) were strongly indicated in 2002 owing to analyses of heart-transplant cases with gender mismatched donors [89] and since then cardiac stem cells have been identified [see [90] and [91] for review]. As with the marrow mesenchymal stem cells the cardiac stem cells are recognized as a heterogeneous population with varying plasticity and growth kinetics.

Not only cardiac but also skeletal muscle contains a number of resident multipotent cell populations. Satellite cells found under the muscle basal lamina, endomysial cells found in the connective tissue sheath surrounding muscle fibers and pericytes associated with the microvasculature are three of several cell types found in human skeletal muscle that show a remarkable degree of plasticity *in vitro* [92]. Even primary skeletal muscle cells have been shown to readily differentiate to smooth muscle cells under specific conditions *in vitro,* supporting capillary network formation in co-culture with endothelial cells [93]. The satellite cells and endomysial cells readily differentiate along the myogenic lineage to produce myoblasts, but further studies have shown plasticity that supports adipogenesis, chondrogenesis and osteogenesis under specific conditions. A substantial portion of research regarding muscle cells and their differentiation capacities revolve around exploring therapy options for pathologies of muscle tissue, such as Duchenne muscular dystrophy, as well as studies focused on understanding the cellular and molecular background of such pathologies.

The pericytes, found on capillary walls throughout the body, show similar plasticity as marrow mesenchymal stem cells, but have been suggested to be mesenchymal progenies of the embryonic mesoangioblast which may account for their localization and surface marker profile [94]. Pericytes were discovered already in 1873 by Rouget, and since then at least three variants have been distinguished histologically. Even these are recognized as being heterogeneous. The pericytes are studied in their major roles in blood-brain barrier maintenance, angiogenesis and neovascularization. These cells seem to be able to function both as maintainers of dynamic microenvironments as well as respond with phenotype alteration both *in vivo* and *in vitro*. [These aspects are reviewed in [95]]. Interesting multipotent cells are also found in the circulation and merit mention. An interesting observation was published in 1994 by Bucala *et al.* [96] describing the use of a murine model for studying the early wound healing response. The influx of blood borne cells (neutrophils, monocytes and lymphocytes) to the wound was accompanied by a large number of adherent, spindle-shaped cells reminiscent of fibroblasts. The migration of fibroblasts from adjacent stroma is common in wound healing but, in the view of the authors, the migration alone could not explain the extent and speed of the fibroblast repopulation. Instead, a cell type originating from the circulation was postulated. Applying gradient centrifugation and plastic adhesion selection on blood cells resulted in the isolation of a fibroblastic cell type. It was characterized as positive for collagen I and III, vimentin, fibronectin, CD11b, CD18, CD45,

CD34 and CD71. The morphology and marker profile as well as function showed similarities with mesenchymal stem cells, the hematopoietic lineage and embryonic fibroblasts; as such, the cell type was termed "fibrocyte" [96]. At least two versions of fibrocytes have recently been described, either differentiated *in vitro* from CD14-positive monocytes or from peripheral blood mononuclear cells [97]. Fibrocytes can differentiate to bronchial myofibroblasts [98], and adipocytes [99] and are believed to be an important indicator of disease under some conditions [reviewed in [100]. Much of the interest in fibrocytes involves their role in fibrosis in the lungs and other organs, and their roles in supporting healthy and diseased microenvironments. Still, more research into fibrocyte biology is needed to determine their use in tissue engineering and regenerative medicine.

There are descriptions of other circulating populations of multipotent cells, such as the endothelial progenitor cells [101], and the term "circulating stem cell" appears in several publications. The clinical emphasis of many of these studies complicates a consensus view on the origin, nature and relation of these circulating cells. These cells are not yet widely explored for suitability in tissue engineering despite reports of interesting plasticity and feasible isolation.

In contrast to the as yet largely undefined circulating stem cells, the skin hosts a number of reported stem cell populations. The skin is the body's largest organ and contains many diverse structures, including a plethora of cell types. The outermost layer, the epidermis, is composed of keratinocytes that originate from progenitors in the basal lamina. The dermis is found beneath the basal lamina, composed of multiple structures including hair follicles, sebaceous glands, sweat glands and microvasculature. Beneath the dermis is the subcutaneous adipose tissue. The subcutaneous layer is rich in preadipocytes, and the microvasculature in the skin contains pericytes and fibrocytes. Stem cells residing adjacent to the sebaceous gland in the hair follicles demonstrate a notable degree of plasticity and have been considered for use in tissue engineering applications. The major focus of research however, is concerned with using these cells in hair regeneration and cell therapy combinations [reviewed by Cotsarelis in [102]]. Their neuroectodermal origins provide an interesting background and diverse plasticity compared to most other types of multipotent cells mentioned thus far and some groups hope to use these in neuronal regenerative therapies [compared and reviewed in [103]].

While the hair bulge contains a population of hair follicle stem cells, other stem cells and/or multipotent cells reside throughout the dermis. Toma and co-workers have presented a way to isolate what they term skin-derived precursors, a population believed to originate from ectoderm-determined embryonal tissue that persists in adult skin [extensively reviewed by Toma *et al.* in [104]]. These skin-derived precursors have been shown to reside in the dermal papillae of hair follicles and are considered pluripotent [105].

Adipose tissue, abundant subcutaneously, contains a readily available store of multipotent cells - the preadipocytes. The immortalized 3T3-L1 cell line has been used extensively for research into adipocyte differentiation, providing useful insights into molecular mechanisms of phenotype control [recently reviewed in [106]. Use of the 3T3-L1 model is not always suitable for adipocyte research but primary human preadipocytes can be isolated with relative ease and have hence become common for *in vitro* studies. Described by Zuk in 2001 the preadipocytes, obtained from lipoaspirate, were shown to support adipogenesis, chondrogenesis, myogenesis and osteogenesis *in vitro*, qualifying them as a feasible source for tissue engineering applications [107]. Preadipocytes, or adipose-derived stem cells, have additionally been shown to differentiate into both ectodermal and endodermal phenotypes [reviewed in [108] and [109]]. Along with mesenchymal stem cells the preadipocytes are regarded as suitable cells for use in tissue engineering applications although some may view these cells as demonstrating a higher degree of commitment and therefore a lower degree of plasticity than mesenchymal stem cells.

The skin has proven to be a rich environment supporting multiple cell types that demonstrate multi-lineage potential. The most prominent cell in the dermis, however, is the fibroblast. Fibroblasts are a very heterogeneous population responsible for the maintenance of connective tissues throughout the body. Dermal fibroblasts from different anatomical sites show differences in gene expression patterns, and fibroblasts from different tissue stroma can show large variations in surface marker profiles [reviewed by Sorrell and Caplan in [110] and [111]]. As Friedenstein found, the fibroblasts have a large role in maintaining suitable stromal environments to support the physiological roles of each specific tissue, and this can motivate the phenotypic variations found within the group of cells called fibroblasts. Reports of multi-lineage induction of human dermal fibroblasts for tissue engineering purposes show that sub-sets of fibroblasts, indistinguishable by common mesenchymal markers, hold similar plasticity as preadipocytes and show a spontaneous heterogeneity within clonal populations [112].

There are at least two things that are certain when considering multipotent cell sources for tissue engineering. Firstly, not enough is known about any phenotype to fully characterize any cell type at this time - there is still too much that we do not know concerning genetic regulation. Secondly, multipotent cells can be found in virtually all compartments of the adult human body, and these can be expanded and differentiated under appropriate conditions *in vitro*, whether they are stem cells or not. Altogether, the natural sources of phenotypic plasticity are only still beginning to be understood whilst our biological proficiency is opening new doors to using our own cells for tissue engineering.

5.3. *In Vitro* Differentiation Update: Media Supplementation and Transfection

Control of cells outside of the body is needed for the goal of in vitro engineered tissue to be successful. Fundamentally it is believed that the in vitro environment must mimic the in vivo niche to maintain cells in a functional and relevant state. To date, a large portion of adherent cell culture is performed in cell culture plastic flasks in the presence of a version of Eagle's medium, such as Dulbecco's Modified Eagle's Medium (DMEM), and bovine serum. Harry Eagle, working to formulate his own optimal cell culture medium in 1955, understood that there are numerous factors present in the body that influence cell behavior and growth kinetics, and that these are needed in specific concentrations for successful cell cultures [113]. Cells are capable of responding to numerous growth factors that can be present in tissue, and these often have radical effects on cell behavior. Adding such factors to cells in culture is a common method of causing changes to the overall phenotype distribution. This medium-supplementation scheme is readily used in tissue engineering to differentiate multipotent cells. The triad of differentiation stimuli commonly used to demonstrate mesodermal multipotency is adipogenic, chondrogenic and osteogenic induction, although many studies include myogenic differentiation as well. The stated reasoning behind these experiments is usually to prove a capacity of the given cell type to differentiate along multiple lineages, thereby showing stem cell plasticity. This is something that is deemed important for a cell source in tissue engineering as it is one part of what makes up an ideal source of cells: cells should be convenient to harvest, readily available, easy to control in culture, be autologous, and have multilineage potential.

By now there are many reports of multilineage inducible cell types. Some types were mentioned in the previous section, but just as the classification of a homogeneous populations seems to break apart when more detailed analyses are applied, so uncertainty about the target phenotypes are revealed. Surprisingly, analyses that support a confirmation of differentiation do so only inasmuch as assumptions are not disproven. In the case of adipogenesis, to mention an example, the accumulation of cytoplasmic inclusions is a strong indicator of fat cell maturation as these are expected to be lipid vesicles. Lipophilic dye, commonly Oil Red O, is used to stain these vesicles and as the lipid content is considered proven, so is the adipogenic nature of the cell. The implicit assumption is that the presenting of a feature considered specific for a given phenotype, in this case lipid droplets for adipocytes, is enough to establish the phenotypic identity of the cells. The weakness in this approach is much to blame on the lack of knowledge of how the lipid vesicles come about at the gene and protein levels. We are yet to understand whether many cells are in fact capable of creating lipid vesicles but rarely do so on account of lack of specific stimuli, or whether it is in fact a unique feature of the adipocyte phenotype. Multiple analyses are usually employed to alleviate the uncertainty of phenotype characterization. To continue the adipocyte example; if lipid vesicles are found and expression of adipocyte-typic proteins and genes is shown then the evidence makes a stronger case for adipogenesis. After this level of detail two scientific approaches diverge: the clinical perspective dictates that the identity of the target cell is less important than its functionality in a given setting, whereas the interest in basic sciences lies as much in determining where and why the differences between the target cell and the achieved cell appear. Today, to prove survival and integration of cells in vivo is the most sought after approach considering the clinical tendency of tissue engineering research. Most of the deeper analyses of cell differentiation biology are carried out outside of the tissue engineering field, but findings can have direct effects on how cells are used in tissue engineering.

The approach of transfection, by Takahashi and Yamanaka to create induced pluripotent stem cells [73, 74], has provided an exciting alternative to medium supplementation. The advent of induced pluripotent stem cells (iPSC) opened up a new way of looking at cellular engineering and what might be possible to achieve with genetic techniques. In the case of iPSC generation the goal is to "revert" cells to a primitive pluripotent state, but using the same techniques with different factors it is possible to perform direct phenotypic conversions instead. Such approaches has been used to obtain neurons [116], cardiomyocytes [117], macrophage-like cells [118] and blood progenitors [119] from fibroblasts. By using genetic techniques specific genes can be added or silenced, thereby modifying the transcriptome quite directly. The approach has received criticism for its lack of clinical suitability on account of the widespread use of viral vectors and genome modifications, but alternative approaches have surfaced that alleviate these criticisms. The use of human recombinant proteins with poly-arginine conjugations has been tried as an alternative to transfection [114]. This approach is a return to medium supplementation, but as published it was less efficient and more expensive than similar viral techniques. Another approach is supplementing the medium with modified mRNA. This has been shown to be relatively efficient compared to traditional iPSC generation and also avoids modifying the cells' genome [115]. The process of selecting the factors for direct conversion is still not generalized enough that it can be done for any given cell type, but the variation of applicable techniques along with the continuing research on phenotypic characterizations will undoubtedly prove immensely useful for tissue engineering.

CONCLUSION

As the medical and biological sciences become increasingly complex, multidisciplinary approaches are needed to make progress. This is evident in the case of tissue engineering. To determine the complex interplay between transplanted cells, biomaterials and host tissue, detailed knowledge of both cellular biology and materials science is indispensable.

In this chapter, a couple of key disciplines have been introduced and exemplified through the efforts to control and promote nerve regeneration. Here, a better understanding of the healing process is necessary and we still lack knowledge of the roadmap to achieve this in full. Furthermore, we need to study how we can use materials to deliver the required therapeutic agents, be they cell, drug or gene deliveries, in an efficient way.

There is a plethora of human multipotent cell sources, readily available for tissue engineering applications. Fundamental questions regarding the biological mechanisms underlying cell identity and function are still unanswered, leading to uncertainty over stability and controllability of tissue engineered constructs. This is echoed in the judiciousness with which tissue engineering applications are approaching the clinic, but also spurs the progress of research within tissue engineering.

While young and still immature, tissue engineering is an exciting field of research that holds great promise and constant progress is being made within it.

REFERENCES

[1] Thoreau, H., Walden; or, *Life in the Woods.* 1854: Ticknor and Fields: Boston.
[2] Nerem, R.M., Tissue engineering: the hope, the hype, and the future. *Tissue Eng.,* 2006. 12(5): p. 1143-50.
[3] Hayflick, L. and P.S. Moorhead, The serial cultivation of human diploid cell strains. *Exp. Cell Res.,* 1961. 25: p. 585-621.
[4] Goldberg, A.D., C.D. Allis, and E. Bernstein, Epigenetics: a landscape takes shape. *Cell,* 2007. 128(4): p. 635-8.
[5] Goldberg, M., R. Langer, and X.Q. Jia, Nanostructured materials for applications in drug delivery and tissue engineering. *Journal of Biomaterials Science-Polymer Edition,* 2007. 18(3): p. 241-268.
[6] Stevens, M.M. and J.H. George, Exploring and engineering the cell surface interface. *Science,* 2005. 310(5751): p. 1135-1138.
[7] O'Brien, F.J., et al., The effect of pore size on cell adhesion in collagen-GAG scaffolds. *Biomaterials,* 2005. 26(4): p. 433-441.

[8] Baker, S.C., et al., The relationship between the mechanical properties and cell behaviour on PLGA and PCL scaffolds for bladder tissue engineering. *Biomaterials,* 2009. 30(7): p. 1321-1328.

[9] Leipzig, N.D. and M.S. Shoichet, The effect of substrate stiffnesson adult neural stem cell behavior. *Biomaterials,* 2009. 30(36): p. 6867-6878.

[10] Dado, D. and S. Levenberg, Cell-scaffold mechanical interplay within engineered tissue. *Seminars in Cell and Developmental Biology*, 2009. 20(6): p. 656-664.

[11] Gilbert, P.M., et al., Substrate Elasticity Regulates Skeletal Muscle Stem Cell Self-Renewal in Culture. *Science*, 2010: p. science.1191035.

[12] Christopherson, G.T., H. Song, and H.Q. Mao, The influence of fiber diameter of electrospun substrates on neural stem cell differentiation and proliferation. *Biomaterials,* 2009. 30(4): p. 556-564.

[13] Straley, K.S., C.W.P. Foo, and S.C. Heilshorn, Biomaterial Design Strategies for the Treatment of Spinal Cord Injuries. *Journal of Neurotrauma*, 2010. 27(1): p. 1-19.

[14] Lim, S.H. and H.Q. Mao, Electrospun scaffolds for stem cell engineering. *Advanced Drug Delivery Reviews*, 2009. 61(12): p.1084-1096.

[15] Chua, K.N., et al., Surface-aminated electrospun nanofibers enhance adhesion and expansion of human umbilical cord blood hematopoietic stem/progenitor cells. *Biomaterials*, 2006. 27(36): p. 6043-6051.

[16] Chua, K.N., et al., Functional nanofiber scaffolds with different spacers modulate adhesion and expansion of cryopreserved umbilical cord blood hematopoietic stem/progenitor cells. *Experimental Hematology*, 2007. 35(5): p. 771-781.

[17] Nisbet, D.R., et al., Neurite infiltration and cellular response to electrospun polycaprolactone scaffolds implanted into the brain. *Biomaterials,* 2009. 30(27): p. 4573-4580.

[18] Lee, S.J., et al., The effect of surface wettability on induction and growth of neurites from the PC-12 cell on a polymer surface. *Journal of Colloid and Interface Science,* 2003. 259(2): p. 228-235.

[19] Li, B., et al., Influence of carboxyl group density on neuron cell attachment and differentiation behavior: Gradient-guided neurite outgrowth. *Biomaterials,* 2005. 26(24): p. 4956-4963.

[20] Howard, D., et al., Tissue engineering: strategies, stem cells and scaffolds. *Journal of Anatomy,* 2008. 213(1): p. 66-72.

[21] Foo, C., et al., Two-component protein-engineered physical hydrogels for cell encapsulation. *Proceedings of the National Academy of Sciences of the United States of America,* 2009. 106(52): p. 22067-22072.

[22] Place, E.S., et al., Synthetic polymer scaffolds for tissue engineering. *Chemical Society Reviews,* 2009. 38(4): p. 1139-1151.

[23] Ariga, K. and T. Kunitake, *Supramolecular Chemistry - Fundamentals ans Applications*. 1st ed. 2006, Heidelberg: Springer-Verlag.

[24] Li, D. and Y.N. Xia, Electrospinning of nanofibers: Reinventing the wheel? *Advanced Materials,* 2004. 16(14): p. 1151-1170.

[25] Gupta, B., Electrospun Core Sheet Fibers for Soft Tissue Engineering. *National Textile Center Annual Report*, 2005. Project No: F05-NS04:p. 1-6.

[26] Boudriot, U., et al., Electrospinning approaches toward scaffold engineering - A brief overview. *Artificial Organs,* 2006. 30(10): p. 785-792.

[27] Schiffman, J.D. and C.L. Schauer, A review: Electrospinning of biopolymer nanofibers and their applications. *Polymer Reviews,* 2008. 48(2): p. 317-352.

[28] Badrossamay, M.R., et al., Nanofiber Assembly by Rotary Jet-Spinning. *Nano Letters,* 2010. 10(6): p. 2257-2261.

[29] Ruiz-Hitzky, E., et al., Advances in Biomimetic and Nanostructured Biohybrid Materials. *Advanced Materials*, 2010. 22(3): p. 323-336.

[30] Vallet-Regi, M., Nanostructured mesoporous silica matrices in nanomedicine. *Journal of Internal Medicine,* 2010. 267(1): p. 22-43.

[31] Naff, N.J. and J.M. Ecklund, History of peripheral nerve surgery techniques. *Neurosurgery clinics of North America*, 2001. 12(1): p. 197.

[32] Mackinnon, S.E., A.L. Dellon, and J. Zhu, *Surgery of the peripheral nerve.* 1988: Thieme Medical Publishers New York.

[33] Norkus, T., M. Norkus, and T. Ramanauskas, Donor, recipient and nerve grafts in brachial plexus reconstruction: anatomical and technical features for facilitating the exposure. *Surgical and Radiologic Anatomy,* 2005. 27(6): p. 524-530.

[34] Terzis, J., B. Faibisoff, and H.B. Williams, The nerve gap: suture under tension vs. graft. *Plastic and Reconstructive Surgery*, 1975. 56(2):p. 166.

[35] Kotani, T., et al., Postoperative results of nerve transposition in brachial plexus injury. Seikeigeka. *Orthopedic surgery*, 1971. 22(11): p. 963.

[36] Tuttle, H.K., Exposure of the brachial plexus with nerve-transplantation. *Jama,* 1913. 61(1): p. 15.

[37] Brandt, K.E. and S.E. Mackinnon, A technique for maximizing biceps recovery in brachial plexus reconstruction. *The Journal of hand surgery,* 1993. 18(4): p. 726-733.

[38] El-Gammal, T.A. and N.A. Fathi, Outcomes of surgical treatment of brachial plexus injuries using nerve grafting and nerve transfers. *Journal of reconstructive microsurgery,* 2002. 18(1): p. 7-14.

[39] Samardzic, M., et al., Transfer of the medial pectoral nerve: Myth or reality? *Neurosurgery,* 2002. 50(6): p. 1277.

[40] Waikakul, S., S. Wongtragul, and V. Vanadurongwan, Restoration of Elbow Flexion in Brachial Plexus Avulsion Injury: Comparing Spinal Accessory Nerve Transfer With Intercostal Nerve Transfer* 1.*The Journal of hand surgery,* 1999. 24(3): p. 571-577.

[41] Ballance, C.A., H.A. Ballance, and P. Stewart, Remarks on the operative treatment of chronic facial palsy of peripheral origin. *British Medical Journal*, 1903. 1(2209): p. 1009.

[42] Harris, W. and V.W. Low, On the importance of accurate muscular analysis in lesions of the brachial plexus; and the treatment of Erb's palsy and infantile paralysis of the upper extremity by cross-union of the nerve roots. *The British Medical Journal,* 1903. 2(2234): p. 1035-1038.

[43] Viterbo, F., et al., End-to-side neurorrhaphy with removal of the epineurial sheath: an experimental study in rats. *Plastic and reconstructive surgery*, 1994. 94(7): p. 1038.

[44] Amr, S.M. and A.N. Moharram, Repair of brachial plexus lesions by end-to-side side-to-side grafting neurorrhaphy: experience based on 11 cases. *Microsurgery,* 2005. 25(2): p. 126-146.

[45] Lundborg, G., et al., Can sensory and motor collateral sprouting be induced from intact peripheral nerve by end-to-side anastomosis? *The Journal of Hand Surgery: Journal of the British Society for Surgery of the Hand*, 1994. 19(3): p. 277-282.

[46] Luo, Y., T. Wang, and H. Fang, Preliminary investigation of treatment of ulnar nerve defect by end-to-side neurorrhaphy. Zhongguo xiu fu chong jian wai ke za zhi= Zhongguo xiufu chongjian waike zazhi= *Chinese journal of reparative and reconstructive surgery*, 1997. 11(6): p. 338.

[47] Tham, S.K.Y. and W.A. Morrison, Motor collateral sprouting through an end-to-side nerve repair***. *The Journal of hand surgery*, 1998. 23(5): p. 844-851.

[48] Yan, J.G., et al., A modified end-to-side method for peripheral nerve repair: large epineurial window helicoid technique versus small epineurial window standard end-to-side technique. *The Journal of hand surgery,* 2002. 27(3): p. 484-492.

[49] Zhao, J.Z., Z.W. Chen, and T.Y. Chen, Nerve regeneration after terminolateral neurorrhaphy: experimental study in rats. *Journal of reconstructive microsurgery,* 1997. 13(1): p. 31-38.

[50] Bertelli, J.A. and M.F. Ghizoni, Nerve repair by end-to-side coaptation or fascicular transfer: a clinical study. *Journal of reconstructive microsurgery,* 2003. 19(5): p. 313-318.

[51] Bertelli, J.A. and M.F. Ghizoni, Concepts of nerve regeneration and repair applied to brachial plexus reconstruction. *Microsurgery*, 2006. 26(4): p. 230-244.

[52] Hayashi, A., et al., Collateral sprouting occurs following end-to-side neurorrhaphy. *Plastic and reconstructive surgery*, 2004. 114(1): p. 129.

[53] Hayashi, A., et al., Axotomy or compression is required for axonal sprouting following end-to-side neurorrhaphy. *Experimental neurology*, 2008. 211(2): p. 539-550.

[54] Matsumoto, M., et al., Schwann cells can induce collateral sprouting from intact axons: experimental study of end-to-side neurorrhaphy using a Y-chamber model. *J. Reconstr. Microsurg.,* 1999. 15(4): p. 281-286.

[55] Evans, P.J., et al., Regeneration across preserved peripheral nerve grafts. *Muscle and nerve*, 1995. 18(10): p. 1128-1138.

[56] Lawson, G.M. and M.A. Glasby, A comparison of immediate and delayed nerve repair using autologous freeze-thawed muscle grafts in a large animal model: The simple injury. The Journal of Hand Surgery: *Journal of the British Society for Surgery of the Hand,* 1995. 20(5): p. 663-670.

[57] Gulati, A.K., Immune response and neurotrophic factor interactions in peripheral nerve transplants. *Acta haematologica*, 1998. 99(3): p. 171.

[58] Trumble, T.E. and F.G. Shon, The physiology of nerve transplantation. *Hand clinics,* 2000. 16(1): p. 105.

[59] Moore, A.M., et al., Limitations of conduits in peripheral nerve repairs. *Hand*, 2009. 4(2): p. 180-186.

[60] Rowland, J.W., et al., Current status of acute spinal cord injury pathophysiology and emerging therapies: promise on the horizon. *Neurosurgical Focus*, 2008. 25(5).

[61] Fleming, J.C., et al., The cellular inflammatory response in human spinal cords after injury. *Brain*, 2006. 129: p. 3249-3269.

[62] Laird, M.D., J.R. Vender, and K.M. Dhandapani, Opposing roles for reactive astrocytes following traumatic brain injury. *Neurosignals*, 2008. 16(2-3): p. 154-164.

[63] Waxman, S., *Molecular Neurology*. 2007, London: Elsevier Academic Press.

[64] Fitch, M.T., et al., Cellular and molecular mechanisms of glial scarring and progressive cavitation: In vivo and in vitro analysis of inflammation-induced secondary injury after CNS trauma. *Journal of Neuroscience*, 1999. 19(19): p. 8182-8198.

[65] Rossi, S.L. and H.S. Keirstead, Stem cells and spinal cord regeneration. *Current Opinion in Biotechnology,* 2009. 20(5): p. 552-562.

[66] Raineteau, O., Plastic responses to spinal cord injury. *Behavioural Brain Research,* 2008. 192(1): p. 114-123.

[67] David, S. and S. Lacroix, Molecular approaches to spinal cord repair. *Annual Review of Neuroscience,* 2003. 26: p. 411-440.

[68] Ropper, A.H. and M.A. Samuels, *Adams and Victor's Principles of Neurology*. 2009, The McGraw-Hill Companies, Inc.: New York.

[69] Raisman, G. and Y. Li, Repair of neural pathways by olfactory ensheathing cells. *Nature Reviews Neuroscience*, 2007. 8(4): p. 312-319.

[70] Oudega, M. and X.M. Xu, Schwann cell transplantation for repair of the adult spinal cord. *Journal of Neurotrauma*, 2006. 23(3-4): p. 453-467.

[71] Enzmann, G.U., et al., Functional considerations of stem cell transplantation therapy for spinal cord repair. *Journal of Neurotrauma*, 2006. 23(3-4): p. 479-495.

[72] Barnabé-Heider, F. and J. Frisén, Stem Cells for Spinal Cord Repair. *Cell Stem Cell,* 2008. 3(1): p. 16-24.

[73] Takahashi, K., et al., Induction of pluripotent stem cells from adult human fibroblasts by defined factors. *Cell,* 2007. 131(5): p. 861-872.

[74] Takahashi, K. and S. Yamanaka, Induction of pluripotent stem cells from mouse embryonic and adult fibroblast cultures by defined factors. *Cell*, 2006. 126(4): p. 663-676.

[75] Barnabé-Heider, F., et al., Origin of New Glial Cells in Intact and Injured Adult Spinal Cord. *Cell Stem Cell*, 2010. 7(4): p. 470-482.

[76] Moroni, L., J.R. De Wijn, and C.A. Van Blitterswijk, Integrating novel technologies to fabricate smart scaffolds. *Journal of Biomaterials Science-Polymer Edition*, 2008. 19(5): p. 543-572.

[77] Tsai, E.C., et al., Synthetic hydrogel guidance channels facilitate regeneration of adult rat brainstem motor axons after complete spinal cord transection. *Journal of Neurotrauma,* 2004. 21(6): p. 789-804.

[78] Tsai, E.C., et al., Matrix inclusion within synthetic hydrogel guidance channels improves specific supraspinal and local axonal regeneration after complete spinal cord transection. *Biomaterials,* 2006. 27(3): p. 519-533.

[79] Balgude, A.P., et al., Agarose gel stiffness determines rate of DRG neurite extension in 3D cultures. *Biomaterials,* 2001. 22(10): p. 1077-1084.

[80] Flanagan, L.A., et al., Neurite branching on deformable substrates. *Neuroreport,* 2002. 13(18): p. 2411-2415.

[81] Georges, P.C., et al., Matrices with compliance comparable to that of brain tissue select neuronal over glial growth in mixed cortical cultures. *Biophysical Journal,* 2006. 90(8): p. 3012-3018.

[82] Saha, K., et al., Substrate Modulus Directs Neural Stem Cell Behavior. *Biophysical Journal,* 2008. 95(9): p. 4426-4438.

[83] Lim, S.H., et al., The effect of nanofiber-guided cell alignment on the preferential differentiation of neural stem cells. *Biomaterials,* 2010. 31(34): p. 9031-9039.

[84] Friedenstein, A.J., et al., Precursors for fibroblasts in different populations of hematopoietic cells as detected by the in vitro colony assay method. *Exp. Hematol.,* 1974. 2(2): p. 83-92.

[85] Friedenstein, A., Stromal-hematopoietic interrelationships: Maximov's ideas and modern models. *Haematol. Blood Transfus*, 1989. 32: p.159-67.

[86] Kuznetsov, S.A., A.J. Friedenstein, and P.G. Robey, Factors required for bone marrow stromal fibroblast colony formation in vitro. *Br. J. Haematol.*, 1997. 97(3): p. 561-70.

[87] Pittenger, M.F., et al., Multilineage potential of adult human mesenchymal stem cells. *Science*, 1999. 284(5411): p. 143-7.

[88] Boheler, K.R., Functional markers and the "homogeneity" of human mesenchymal stem cells. *J. Physiol.,* 2004. 554(Pt 3): p. 592.

[89] Quaini, F., et al., Chimerism of the transplanted heart. *N. Engl. J. Med.,* 2002. 346(1): p. 5-15.

[90] Torella, D., et al., Resident human cardiac stem cells: role in cardiac cellular homeostasis and potential for myocardial regeneration. *Nat. Clin. Pract. Cardiovasc. Med.*, 2006. 3 Suppl 1: p. S8-13.

[91] Barile, L., et al., Endogenous cardiac stem cells. *Prog. Cardiovasc. Dis.*, 2007. 50(1): p. 31-48.

[92] Lecourt, S., et al., Characterization of distinct mesenchymal-like cell populations from human skeletal muscle in situ and in vitro. *Exp. Cell Res.*, 2010. 316(15): p. 2513-26.

[93] Le Ricousse-Roussanne, S., et al., Ex vivo generation of mature and functional human smooth muscle cells differentiated from skeletal myoblasts. *Exp. Cell Res.*, 2007. 313(7): p. 1337-46.

[94] Dellavalle, A., et al., Pericytes of human skeletal muscle are myogenic precursors distinct from satellite cells. *Nat. Cell Biol.,* 2007. 9(3): p. 255-67.

[95] Krueger, M. and I. Bechmann, CNS pericytes: concepts, misconceptions, and a way out. *Glia,* 2010. 58(1): p. 1-10.

[96] Bucala, R., et al., Circulating fibrocytes define a new leukocyte subpopulation that mediates tissue repair. *Mol. Med.,* 1994. 1(1):p. 71-81.

[97] Curnow, S.J., et al., Distinct types of fibrocyte can differentiate from mononuclear cells in the presence and absence of serum. *PLoS One*, 2010. 5(3): p. e9730.

[98] Schmidt, M., et al., Identification of circulating fibrocytes as precursors of bronchial myofibroblasts in asthma. *J. Immunol.*, 2003. 171(1): p. 380-9.

[99] Hong, K.M., et al., Characterization of human fibrocytes as circulating adipocyte progenitors and the formation of human adipose tissue in SCID mice. *FASEB J.,* 2005. 19(14): p. 2029-31.

[100] Herzog, E.L. and R. Bucala, Fibrocytes in health and disease. *Exp. Hematol.,* 2010. 38(7): p. 548-56.

[101] Schmidt-Lucke, C., et al., Quantification of circulating endothelial progenitor cells using the modified ISHAGE protocol. *PLoS One*, 2010. 5(11): p. e13790.

[102] Yang, C.C. and G. Cotsarelis, Review of hair follicle dermal cells. *J. Dermatol. Sci.,* 2010. 57(1): p. 2-11.

[103] Amoh, Y., K. Katsuoka, and R.M. Hoffman, The advantages of hair follicle pluripotent stem cells over embryonic stem cells and induced pluripotent stem cells for regenerative medicine. *J. Dermatol. Sci.*, 2010. 60(3): p. 131-7.

[104] Fernandes, K.J., J.G. Toma, and F.D. Miller, Multipotent skin-derived precursors: adult neural crest-related precursors with therapeutic potential. *Philos. Trans R. Soc. Lond. B Biol. Sci.*, 2008. 363(1489): p. 185-98.

[105] Fernandes, K.J., et al., A dermal niche for multipotent adult skin-derived precursor cells. *Nat. Cell Biol.,* 2004. 6(11): p. 1082-93.

[106] Poulos, S.P., M.V. Dodson, and G.J. Hausman, Cell line models for differentiation: preadipocytes and adipocytes. *Exp. Biol. Med.* (Maywood), 2010. 235(10): p. 1185-93.

[107] Zuk, P.A., et al., Multilineage cells from human adipose tissue: implications for cell-based therapies. *Tissue Eng.,* 2001. 7(2): p. 211-28.

[108] Mizuno, H., Adipose-derived stem and stromal cells for cell-based therapy: current status of preclinical studies and clinical trials. *Curr. Opin. Mol. Ther.,* 2010. 12(4): p. 442-9.

[109] Rada, T., R.L. Reis, and M.E. Gomes, Adipose tissue-derived stem cells and their application in bone and cartilage tissue engineering. *Tissue Eng. Part B Rev.*, 2009. 15(2): p. 113-25.

[110] Sorrell, J.M. and A.I. Caplan, Fibroblasts-a diverse population at the center of it all. *Int. Rev. Cell Mol. Biol.*, 2009. 276: p. 161-214.

[111] Sorrell, J.M. and A.I. Caplan, Fibroblast heterogeneity: more than skin deep. *J. Cell Sci.*, 2004. 117(Pt 5): p. 667-75.

[112] Junker, J.P., et al., Adipogenic, chondrogenic and osteogenic differentiation of clonally derived human dermal fibroblasts. *Cells Tissues Organs*, 2010. 191(2): p. 105-18.

[113] Eagle, H., Nutrition needs of mammalian cells in tissue culture. *Science,* 1955. 122(3168): p. 501-14.

[114] Yang, W.C., et al., Cell-free production of transducible transcription factors for nuclear reprogramming. *Biotechnol. Bioeng.*, 2009. 104(6): p. 1047-58.

[115] Warren, L., et al., Highly efficient reprogramming to pluripotency and directed differentiation of human cells with synthetic modified mRNA. *Cell Stem Cell,* 2010. 7(5): p. 618-30.

[116] Vierbuchen, T., et al., Direct conversion of fibroblasts to functional neurons by defined factors. *Nature,* 2010. 463(7284): p. 1035-41.

[117] Ieda, M., et al., Direct reprogramming of fibroblasts into functional cardiomyocytes by defined factors. *Cell*, 2010. 142(3): p. 375-86.

[118] Feng, R., et al., PU.1 and C/EBPalpha/beta convert fibroblasts into macrophage-like cells. *Proc. Natl. Acad. Sci. USA*, 2008. 105(16): p. 6057-62.

[119] Szabo, E., et al., Direct conversion of human fibroblasts to multilineage blood progenitors. *Nature*, 2010. 468(7323): p. 521-6.

INDEX

A

B

C

D

E

F

J

K

L

M

Q

R

T

U

V

W

X

Y

Z